Information Technology in Educational Research and Statistics

Information Technology in Educational Research and Statistics has been co-published simultaneously as *Computers in the Schools*, Volume 15, Numbers 3/4 and Volume 16, Number 1 1999.

Information Technology in Educational Research and Statistics

Leping Liu
D. LaMont Johnson
Cleborne D. Maddux
Editors

Information Technology in Educational Research and Statistics has been co-published simultaneously as *Computers in the Schools*, Volume 15, Numbers 3/4 and Volume 16, Number 1 1999.

CRC Press
Taylor & Francis Group
Boca Raton London New York

CRC Press is an imprint of the
Taylor & Francis Group, an informa business

Information Technology in Educational Research and Statistics has been co-published simultaneously as *Computers in the Schools*, Volume 15, Numbers 3/4 and Volume 16, Number 1 1999.

Reprinted 2009 by CRC Press

Cover design by Thomas J. Mayshock Jr.

The Haworth Press, Inc., 10 Alice Street, Binghamton, NY 13904-1580 USA

Library of Congress Cataloging-in-Publication Data

Information technology in educational research and statistics / Leping Liu, D. LaMont Johnson, Cleborne D. Maddux, editors.

p. cm.

"... Co-published simultaneously as Computers in the schools, volume 15, numbers 3/4 & volume 16, number 1 1999."

Includes bibliographical references and index.

ISBN 0-7890-0958-7 (alk. paper)–ISBN 0-7890-0976-5 (alk. paper)

1. Educational research–Technological innovations. 2. Educational statistics–Technological innovations. 3. Information technology. I. Liu, Leping. II. Johnson, D. LaMont (Dee LaMont), 1939- III. Maddux, Cleborne D., 1942-

LB1028.3.I51945 1999
370′.7′2–dc21

99-051388

Information Technology in Educational Research and Statistics

Contents

INTRODUCTION

Creative Thinking in the Use of Information Technology in Educational Research and Statistics
D. LaMont Johnson
Leping Liu 1

TECHNOLOGY AND RESEARCH

Constructivism: Has the Term Outlived Its Usefulness?
Cleborne D. Maddux
Rhoda Cummings 5

Computers and Qualitative Research
Jerry Willis
Muktha Jost 21

What Influences the Quality of Research: Findings from an Editorial Management Database
Leping Liu
Stuart A. Harrington
D. LaMont Johnson 53

Maintaining Research Documents with Database Management Software
Stuart A. Harrington 67

Using Technology to Promote Teacher Action Research
Alex C. Pan 81

Developing and Researching Information and Communication Technology as a Research Tool
Niki Davis
Roger Trend 101

Software for Analyzing Qualitative Data
Jerry Willis
Muktha Jost 117

Assessing the Impact of Computer-Based Learning: A Research Course for Graduate Instructional Technology Programs
W. Michael Reed 151

TECHNOLOGY AND STATISTICS

A Technology-Based Teaching Model That Stimulates Statistics Learning
George C. J. Fernandez
Leping Liu 173

Computer-Assisted Approach for Teaching Statistical Concepts
Carl Lee 193

The Comparative Effectiveness of Various Microcomputer-Based Software Packages on Statistical Achievement
Edwin Christmann
John Badgett 209

A Web Site as a Tool for Learning Statistics
Christina Dokter
Larry Heimann 221

Index 231

ABOUT THE EDITORS

Leping Liu, PhD, is Assistant Professor of Education in the Department of Reading, Special Education and Instructional Technology at Towson University. She received her PhD in Information Technology in Education and Statistics from the University of Nevada, Reno. She teaches courses on the application of technology in education. Her research interests focus on: integration of technology in classroom teaching/learning, technology and mathematical thinking, multimedia learning environment and cognitive development, and computer achievement modeling. She has authored or co-authored several professional articles and co-edited one book.

D. LaMont Johnson, PhD, is Professor of Education in the Department of Counseling and Educational Psychology at the University of Nevada, Reno. He is also Program Coordinator of the Information Technology in Education program. He teaches courses on the application of technology in education and trains teachers across the state of Nevada in using the Internet in their classrooms. The co-author of the textbook *Educational Computing: Learning with Tomorrow's Technologies*, now in its second edition, Professor Johnson has written or co-written numerous books and articles on educational computing and information technology in education.

Cleborne D. Maddux, PhD, is Professor of Education in the Department of Counseling and Educational Psychology at the University of Nevada, Reno, where he teaches courses on statistics and on integrating technology into education. He trains elementary and high school teachers in the state of Nevada on how to make the Internet a regular feature of their classroom agendas. Senior author of *Educational Computing: Learning with Tomorrow's Technologies*, a textbook now in its second edition, Professor Maddux has authored or co-authored numerous professional articles and books on informational technology in education and educational technology.

INTRODUCTION

D. LaMont Johnson
Leping Liu

Creative Thinking in the Use of Information Technology in Educational Research and Statistics

Educational research has been around for a long time and is based on some traditional practices. At the same time, however, conducting research in education is constantly being influenced by change. Some of this change stems from the fact that there has never been a real sense of satisfaction among educators regarding the capacity of current research methods to provide clear answers to complex questions. Consequently, there is always a search for something new and better. Another reason that educational research is constantly in a state of change is that new tools and techniques are constantly being developed, and those who teach research and do research strive to incorporate these new tools and techniques into the literature and practice of educational research methodology. As with so many fields of study today, we are seeing a rapid proliferation of new tools and techniques. The advent of so many information technology tools in just the past

[Haworth co-indexing entry note]: "Creative Thinking in the Use of Information Technology in Educational Research and Statistics." Johnson, D. LaMont, and Leping Liu. Co-published simultaneously in *Computers in the Schools* (The Haworth Press, Inc.) Vol. 15, No. 3/4, 1999, pp. 1-3; and: *Information Technology in Educational Research and Statistics* (ed: Leping Liu, D. LaMont Johnson, and Cleborne D. Maddux) The Haworth Press, Inc., 1999, pp. 1-3. Single or multiple copies of this article are available for a fee from The Haworth Document Delivery Service [1-800-342-9678, 9:00 a.m. - 5:00 p.m. (EST). E-mail address: getinfo@haworthpressinc.com].

few years is having a profound effect on not only how we do research, but on how we think about research. Chief among these information technology tools is, of course, the Internet and the World Wide Web.

The Internet has influenced educational research in at least two ways. First, our concept of space and time has changed, which affects our desire and willingness to be patient with some of the old traditional research tools and methods. Second, with so much information so readily available, to so many people, it is difficult for authoritative positions to be established or maintained. Thus, we are seeing extreme positions being taken that suggest that research is whatever the researcher says it is. A consequence of such rapid change and upheaval in the field of educational research warrants time and space to reflect on the many new possibilities that are emerging. This special issue focuses on some of these possibilities, particularly as they relate to information technology. Some of the issues our contributors address include on-going debates relating to the theory of research, research methodology, and successful practices.

As we all know, a critical and often-controversial component of educational research is the research design. The fact that good research starts from a good research design and that good research design will produce more meaningful outcomes is a universally supported proposition. While ideally, statistical procedures should be selected to fit the research design, in practice statistical procedures influence research design. Even in its purest form, educational research is influenced by statistics since all quantitative research depends on some form of statistical analysis. Recent trends in educational research mix what is being billed as *quantitative* and *qualitative* research. Such trends certainly call for debate on what statistical procedures are appropriate for what kinds of research design. Some debate on these and related issues is discussed in this issue.

As the power and availability of information technology grow, traditional statistics courses are being modified to take advantage of new software developed for not only carrying out statistical analyses but also for teaching statistical concepts. Some specific software packages are discussed in this issue along with suggestions on their use.

Two trends relating to using technology in educational research emerged from the submissions we received for this issue. The first trend is toward using a variety of software in teaching research. An important question relating to the use of software is, What software should I use? Our contributors, in addressing appropriate software selection, discuss various statistics packages such as *SPSS, SAS, BMD, MyStat, Minitab* for use with quantitative data. They also discuss the use of *TrueStat.* and *NUD*IST, Ethnograph, Kwalitan,* and *ATLAS/ti* for use with qualitative data. Advantages and disadvantages of different software packages are considered. A second question that our contributors consider relating to software selection is, What are effective uses of software in teaching crucial concepts relating to research and statistics? Issues discussed in this regard include technology-based course design, applications of specific software, and teaching/learning methods in certain application environments.

The second trend of using technology in educational research and statistics is related to the ever popular World Wide Web. The Web, of course, is a rich information resource for teaching and research and is being used in numerous ways. Some innovative and creative uses of the Web are suggested:

- Students can use Web information as a source of real-life projects and research data.
- Instructors can use Web sources to conduct interactive demonstrations.
- Instructors can develop online statistics courses, providing course materials, useful links to other online tutorials or demonstrations, and online exams.
- Students and instructors can go to the Web to find debate and analysis on new trends in educational statistics and research.
- Survey research can be conducted online using the Web.

It is the hope of the editors of this special volume that the ideas, debates, and suggestions set forth by a variety of distinguished authors will stimulate the reader to consider new and creative ideas relating to using information technology in educational research and statistics. It is further hoped that the contents of this volume will stimulate further comment and discussion.

TECHNOLOGY AND RESEARCH

Cleborne D. Maddux
Rhoda Cummings

Constructivism: Has the Term Outlived Its Usefulness?

There have been many recent suggestions that traditional research paradigms and educational strategies are in need of abandonment or of radical revision. Some writers suggest, for example, that qualitative research strategies should be given as much credence as quantitative strategies. Others suggest that the qualitative approach should totally replace quantitative methods. Many of these critics maintain that the rationale for this transformation is implicit in a number of diverse philosophical and theoretical ideas such as "postmodernism," "relativism," "phenomenology," "naturalism," "feminism," "neo-marxism," "progressivism," "deconstructivism," and "constructivism."

CLEBORNE D. MADDUX is Professor, Department of Counseling and Educational Psychology, College of Education, University of Nevada-Reno, Reno, NV 89557. E-mail: maddux@unr.edu
RHODA CUMMINGS is Professor, Department of Counseling and Educational Psychology, College of Education, University of Nevada-Reno, Reno, NV 89557. E-mail: cummings@unr.edu

[Haworth co-indexing entry note]: "Constructivism: Has the Term Outlived Its Usefulness?" Maddux, Cleborne D., and Rhoda Cummings. Co-published simultaneously in *Computers in the Schools* (The Haworth Press, Inc.) Vol. 15, No. 3/4, 1999, pp. 5-20; and: *Information Technology in Educational Research and Statistics* (ed: Leping Liu, D. LaMont Johnson, and Cleborne D. Maddux) The Haworth Press, Inc., 1999, pp. 5-20. Single or multiple copies of this article are available for a fee from The Haworth Document Delivery Service [1-800-342-9678, 9:00 a.m. - 5:00 p.m. (EST). E-mail address: getinfo@haworthpressinc.com].

Many of these ideas are sweeping in scope and represent entire fields or subfields of study in various disciplines such as philosophy, sociology, political science, and so forth. Therefore, it is not our intention to address all these topics. Space would not permit, and there are sufficient discussions available elsewhere, particularly discussions of formal philosophical bases (for example, Eisner, 1991; Eisner & Peshkin, 1990; Lancy, 1993).

Instead, we will focus our discussion on the suggestions that have been made for the revision of instructional methods and educational research strategies and the rationale that has been cited in the literature for such changes. Because *constructivism* is the term that seems to be cited most frequently in professional education circles, we will restrict our discussion to the following: (a) we will describe the underlying theoretical foundations of constructivism, (b) we will consider the question of whether it is practical to provide a coherent definition of constructivism, and (c) we will examine whether constructivism has implications for the revision of educational research.

WHAT IS CONSTRUCTIVISM?

"Constructivism" is one of the most popular general topics in recent educational literature, and is especially fashionable in the field of information technology in education. One indication of the general popularity of this topic is that a search of the World Wide Web using the AltaVista search engine and the search string "constructivism" identified more than 17,000 Web sites dealing with the term. Refining the search by adding "education" to the search string still revealed over 6,500 sites, while ERIC lists 1,890 journal articles dealing with both education and constructivism. Furthermore, numerous presentations at national educational conferences over the past several years (such as the yearly international meeting of the Society for Information Technology in Education) have made prominent references to constructivism and its application to various educational specialties, even though there often is no logical rationale for doing so. In addition, the national standards of many professional educational organizations such as the National Council of Teachers of Mathematics and the National Science Teachers Association include numerous references to constructivism. Then, too, our own students in the college of education report that the term *constructivism* is often referred to in their education courses. (However, when we ask them to explain the meaning of the term, most are uncomfortably mute, or respond hesitantly with vague and largely meaningless generalizations.)

One reason students and others have difficulty discussing constructivism is that the term, as it appears in educational literature, is often ill-defined and ambiguous, with little or no agreement about the basic tenets of constructivism. Some authors assert that constructivism is a philosophical explanation (e.g., Airasian & Walsh, 1997), some suggest it is a theory (e.g., Jaramillo, 1996), still others suggest that it is a category into which many theories can be placed (e.g., Slavin, 1997), others assert that it is "a prominent element of many cognitive learning theories" (Ormrod, 1998, p. 206), and still others maintain that it is a teaching approach (e.g., Kindsvatter, Wilen, & Ishler, 1996).

It may seem surprising that educators give so much attention to a term without first agreeing on its meaning. This may be a specific example of what Salman Rushdie has referred to as "a revival in the fine art of meaningless naming. . . . Welcome to the New Incomprehensibility."

It is clear that constructivism is a favorite topic among educators at all levels. However, the term is used to justify a host of contradictory recommendations about everything from school reform, teaching methodologies, and what constitutes educational research and how such research should be conducted. *We believe this occurs because of confusion about the nature and definition of constructivism and failure to read, study, and understand its underlying theoretical foundations.*

As previously mentioned, a host of descriptors for constructivism appear in the literature, including, among others, a *philosophy, an attitude, a methodology, an approach, a model, an epistemology, a framework,* etc. However, the most common descriptor among educators seems to be *theory.* This latter descriptor is part of the problem. *Constructivism is not itself a theory.* It is at best a general term or concept that is loosely and often haphazardly derived from a variety of theories in developmental psychology. As long as constructivism is a poorly understood and variously defined concept among educators, it will remain impossible to apply it in schools or to use it as the basis for educational research.

THE FOUNDATION OF CONSTRUCTIVISM

Although constructivism became a popular term in the early 1990s, its theoretical origins can be traced to much earlier times, but particularly to the early part of the twentieth century. Although scholars and educators attribute constructivism to a number of different theorists, the two most often mentioned are Lev Vygotsky and Jean Piaget. Rather than trying to condense all of the existing definitions into one that is comprehensive and coherent, perhaps it is more useful to examine the basic tenets proposed by these two theorists, especially in terms of how each describes children's cognitive and social development. First, however, it is important to dispel some common misconceptions about these two theorists, especially with regard to similarities and differences in their ideas.

One of the most common errors is the suggestion that their ideas oppose each other, or that Vygotsky's theories disproved those of Piaget. In fact, there are many more similarities than differences inherent in the two theories. Kohlberg and Wertsch (1987), in their discussion of how Piaget and Vygotsky explain the relationship of thought and language, suggest that "overall, we see Piaget and Vygotsky as presenting approaches that complement one another" (p. 219).

The similarities may have been more numerous or the differences greater, however, had Vygotsky's life and career lasted as long as those of Piaget. The two were contemporaries and were even born in the same year (1896). However, Vygotsky died in 1934 at the age of 38, while Piaget lived and continued to work until 1980 when he died at the age of 84.

Piaget and Vygotsky: Basic Tenets

Egocentrism. Because of Vygotsky's premature death from tuberculosis, he was acquainted only with Piaget's earliest work, which was published in the early 1920s (Piaget, 1923; 1924) and which focused on the role of *egocentrism* and its relationship to *self-directed speech* in young children. Both Piaget and Vygotsky noticed that children younger than six or seven years of age often engaged in solitary speech not directed at other people. Such speech took place while the children were engaged in work or play requiring some cognitive task. Both also noticed that the frequency of self-directed speech declined sharply and then seemed to disappear sometime around age 7 or 8.

While they did not disagree about the existence of self-directed speech, Piaget and Vygotsky had different explanations of its role in cognitive development.

According to Piaget (1923, 1924), self-directed speech is egocentric: It reflects the cognitive egocentrism of the young child, which is characterized by an inability to differentiate self from other and a consequent inability to take the perspective of another person. This results in a failure to communicate. As egocentrism diminishes with development, perspective-taking ability increases, communication improves, and self-directed speech declines. Thus, Piaget believed self-directed speech served no useful purpose in development.

Vygotsky disagreed with Piaget's interpretation of self-directed speech as a symptom of egocentric thinking. Instead, he believed that self-directed speech served a valuable problem-solving function in that it directed thinking. To test this idea, Vygotsky designed a series of experiments in which he assigned a task to young children, then increased the difficulty and frustration level of the task. In the more difficult situations, Vygotsky found that children's use of self-directed speech almost doubled. Vygotsky describes one experiment:

> For instance, when a child was getting ready to draw, he would suddenly find that there was no paper, or no pencil of the color he needed. In other words, by obstructing his free activity, we made him face problems. . . . The child would try to grasp and remedy the situation in talking to himself: "Where's the pencil? I need a blue pencil. Never mind, I'll draw with the red one and wet it with water; it will become dark and look like blue." (Vygotsky, 1986, p. 30)

Vygotksy concluded that, as the task difficulty increased for young children, self-directed speech increased; and as difficulty decreased, self-directed speech decreased. He took this as evidence that self-directed speech was not due simply to the child's egocentrism, but was central to thinking processes. Furthermore, like Piaget, Vygotsky noticed that self-directed speech did not commonly occur in school-age children, who tended to be silent while working on cognitive tasks. However, he found that, if such children were asked to think aloud while problem solving, what they verbalized sounded much like the self-directed speech observed in younger children. He took this to mean that the early self-directed speech had become *inner speech* in school-age children, and involved the same thinking processes. According to Vygotksy (1986), "These similarities lead us to assume that when egocentric speech disappears, it does not simply atrophy but 'goes underground,' i.e., turns into inner speech" (p. 33).

In summary, Piaget suggested that self-directed speech was a developmental dead-end that disappeared when the child became less egocentric, gained the ability to take the perspective of others, and developed effective communication skills. Vygotsky, on the other hand, asserted that self-directed speech played an important role in the development of cognition, and was a precursor to inner speech required for effective problem-solving.

Although these differences in the two theories may seem trivial, it must be remembered that Vygotsky, who considered them critical differences, was reacting to the earliest and most narrow of Piaget's descriptions of children's thought processes, and he focused most of his criticism on Piaget's early ideas about egocentrism and concept development in very young children. While these ideas are important in Piaget's theory, they are most important for young children in the *preoperational stage* and are representative of the most embryonic of Piaget's work, the majority of which occurred long after Vygotsky's death.

Socialization in young children. Perhaps a more important difference between the two theories has to do with the fact that Piaget thought it critical to explain how children arrived at a sense of "self" as individual and separate from other things and other people, while Vygotsky focused on explaining how children came to find a place as productive members of various societal groups. Hart, Kohlberg, and Wertsch (1987) point out that these differences can be partially explained by the fact that the Russian language has no word for "self," and, even more importantly, that Vygotsky was working in a unique social-political environment in Russia:

> An essential part of official Soviet policy after the Russian Revolution of 1917 was that the country had to construct a new socialist society in a relatively short time.
>
> . . . In addition to the influence of official Soviet ideology, we would argue that a more deep-seated set of cultural beliefs and values was also behind Vygotsky's and other Soviet psychologists' focus on socialization into the collective, rather than on

> individual self, identity, and so forth. Specifically, it seems to us that the centuries-old Russian tendency to value collective functioning over individual functioning (at least by Western standards) was also behind his and others' approach to socialization. (p. 247)

Thus, for Vygotsky, the social and cultural environment was more central to cognitive development, which proceeded from outside to inside the child. In a sense, Vygotsky suggested that external events produced thinking. For Piaget, the child was born with certain innate cognitive tendencies, the full development of which depended on interaction with the physical and the social environment. For Piaget, the external environment *activated* innate mental processes, but did not *produce* them.

This distinction, however, does not mean that Piaget considered the social environment to be unimportant, another common misconception. In fact, Piaget emphasized that from the moment of birth, the child begins to construct social knowledge based on interactions with people. However, it does seem that Vygotsky and Piaget disagreed about the importance of school instruction in development, as seen in the following quotation by Vygotsky:

> But if Piaget's views on the nature of nonspontaneous concepts were correct, it would follow that such an important factor in the socialization of thought as school learning is unrelated to the inner developmental processes. It would look as if the inner development of the child's thought had no relation to socialization, while socialization had no relevance for the development of the child's concepts. (1986, p. 155)

It seems to us that Piaget might agree with the first of these statements, since he called his ideas a *theory of genetic epistemology*, implying the importance of certain innate human tendencies. It does not follow, however, that he would agree with the second statement concerning the lack of relevance of socialization to concept development. Again and again in his writing, Piaget emphasized the importance of social relationships in bringing about an end to child egocentrism. Here is only one of many of these examples:

> As a matter of fact, we have here, as in the case of child logic, to suppose that social life is necessary to rational development, but

> that it is not sufficient to create the power of reasoning. Without collaboration between his own thought and that of others, the child would not become conscious of the divergences which separate his ego from that of others, and he would take each of his perceptions or conceptions as absolute. . . . Without social life, he would never succeed in understanding the reciprocity of viewpoint, and, consequently, the existence of perspectives, whether geometrical or logical. He would never cease to believe that the sun follows him on his walk. He would be ignorant of the reciprocity of the notions of right and left, of dependence, in short, of relations in general. It is therefore highly probable that the relativity of ideas would elude him. (Piaget, 1995, p. 137)

In summary, both Piaget and Vygotsky considered socialization important to development of the child. However, Vygotsky was most interested in how children's social interactions lead to their development as contributing members of a collective society. Piaget, on the other hand, was more interested in how children's interpersonal experiences led them out of egocentrism eventually to develop rational thought and a sense of individuality and autonomy.

Sequences of Development

Although Piaget's and Vygotsky's ideas differed in a number of ways, they agreed on the importance of development to concept formation. Both recognized qualitative changes in thinking that occurred with maturation. Piaget described these differences as they occurred through a sequence of four stages, beginning in infancy and ending in adolescence. Although Vygotsky did not specify specific stages of cognitive development, he divided the process of concept formation into several stages within three major phases. Regardless of terminology, both viewed cognitive development as movement from the concrete thought of young children to abstract thinking and concept formation beginning in adolescence. Thus, the tenets of both Vygotsky and Piaget indicate that they were true developmental theorists who maintained that abstract thought occurred only late in the developmental sequence and was not accessible to younger children.

This emphasis on development seems to be ignored by many of the current authors who characterize themselves as constructivists. The following by Jaramillo (1996) is a case in point:

> Although one end of the spectrum is delineated by Piaget's notion that we learn in developmentary steps rather than in leaps and bounds, my focus is upon Vygotsky's view of a nonlinear learning sequence, which constitutes the other. Vygotsky's version of constructivism, then, is a non-developmentary view of education whereby a child's intellectual personality and socio-moral knowledge is "constructed" by students internalizing concepts through self-discovery. (p. 135)

When considering the following quote by Vygotsky from *Thought and Language* (1986), it is difficult to see how Jaramillo could have reconciled his conclusion that Vygotsky was not a developmental theorist.

> The development of the processes that eventually result in concept formation begins in earliest childhood, but the intellectual functions that in a specific combination form the psychological basis of the process of concept formation ripen, take shape, and develop only at puberty. . . . It would not be an exaggeration to say that to equate the intellectual operations with three-year-olds with those of adolescents (as some psychologists do) means to use a sort of logic that would deny the existence of sexual maturation in puberty only because certain elements of sexuality are already present in infants. (p. 106)

And elsewhere: "The ascent to concept formation is made in three basic phases, each divided in turn into several stages" (p. 110).

Furthermore, the teaching strategies proposed by many of the modern experts who claim to be constructivists fail to differentiate among methods that are appropriate for children at different developmental levels. A common error, for example, is to propose methods that presuppose abstract conceptual thinking abilities in elementary age children. In fact, many of these proposals about methodology do not even mention grade or age level of students and imply that suggestions are appropriate for all children regardless of age, grade level, or development.

The following are examples of "constructivist" teaching goals and suggestions that have been proposed for students without identifying a target grade or age. These have been taken from a variety of articles (Clements, 1997; Hanley, 1994; Murphy, 1997) and are representative of the constructivist literature:

1. Use cognitive terminology such as *classify, analyze*, and *create* when giving instructions to students.
2. Have students formulate their own hypotheses.
3. Provide and teach children to appreciate multiple perspectives.
4. Encourage self-awareness in knowledge construction.
5. Have students set their own learning goals and objectives.
6. Design activities to encourage meta-cognition (thinking about one's own thinking), self-analysis, self-regulation, self-reflection, and self-awareness.
7. Emphasize problem-solving, higher-order thinking skills, and deep understanding.
8. Develop knowledge through conscious, reflective construction.

Some of the above suggestions are so general and jargon-laden that it is difficult to know what the authors intended, a problem that seems strangely rampant in recent constructivist literature. Others may be appropriate for use with adolescents who are abstract and reflective thinkers, but are not suited for application with younger, more concrete thinkers.

Modern Constructivism and the Neglect of Development

It is difficult to believe that experienced, practicing teachers and other education professionals would propose or endorse a single set of methods for children of all ages. What accounts for this strange phenomenon? We suspect that one cause is that many of these self-proclaimed constructivists are not reading and studying the theorists whose ideas make up the foundation of the movement. Certainly they do not appear to be reading and studying the work of Piaget and Vygotsky, the two theorists who are most often credited with starting constructivism and who emphasized that cognition in young children differed from that of older children and adolescents.

One indication of the lack of study by this movement's seminal thinkers is the widespread use of secondary sources in the constructivist literature. An example from Bodrova and Leong's (1996) *Tools of the Mind*, a book that is subtitled The *Vygotskian Approach to Early Childhood Education*, follows:

> If you are interested in learning more about Vygotsky and his colleagues and students, Van der Veer and Valsnier (1991) and

> Kozulin (1990) give a detailed account of the life of Vygotsky and his ideas, both in and outside of Russia. (p. 6)

It seems to us that a book about Vygotsky should suggest that interested readers consult primary, rather than secondary sources. The attention to secondary sources and the neglect of primary sources is apparent throughout this and many other constructivist articles and books. Of course, neither Piaget nor Vygotsky are "easy reads," and it may be tempting to take others' word for what constitutes these theorists' basic ideas.

Another reason for the neglect of development is that constructivism has become so uncritically fashionable that many educators assume that the simplest and often the most distorted interpretations of its philosophical and theoretical bases are accurate. In other words, simplistic constructivist literature is so popular, so positive, so ubiquitous, and extols goals and objectives that seem so intrinsically good for children, that the movement is highly emotionally appealing to many educators. How can one argue, for example, with the notion that children are not empty and passive vessels to be filled with knowledge by the teacher?

The Uncritical Acceptance of Modern Constructivism

There are a number of reasons for educators' uncritical acceptance of constructivism as it is characterized in recent literature. For one thing, education is well known for its devotion to fashion, although that devotion tends to be fickle as seen in the rise and fall of a number of movements including new mathematics, discovery learning, open education, and assertive discipline, to name only a few. Kauffman (1999), in his discussion of postmodern/deconstructivist philosophy, has said, "We should not underestimate the popularity of nonscientific and aggressively *anti*scientific beliefs today. Nor should we dismiss the danger of such sentiments and frames of mind for education" (p. 249).

Another reason is that reductionist thinking and complex jargon are consistent with the political climate and the conservative mood of the current educational reform agenda. Politicians and policymakers have always been aware that simplistic solutions for complex problems play well in the media: "Violence will end if movies stop portraying violence," "Crime will cease if penalties are more severe," "Learning will improve if we reinvent the schools of our youth." Educators often endorse and promote similar simplistic solutions to our complex educational problems. "Reduced class sizes improve student test scores," "Cooperative learning helps children accept diversity," "Constructivist methods will make students better problem solvers."

Then, too, many modern constructivist writers are fond of setting up "traditional" educational methods as "straw men" to support their advocacy. Examples from the literature are legion. Here is a representative quotation by Hanley (1994):

> The current American classroom, whether grade school or college level, tends to resemble a one-person show with a captive but often comatose audience. Classes are usually driven by "teacher-talk" and depend heavily on textbooks for the structure of the course. There is the idea that there is a fixed world of knowledge that the student must come to know. Information is divided into parts and built into a whole concept. Teachers serve as pipelines and seek to transfer their thoughts and meaning to the passive student. There is little room for student-initiated questions, independent thought, or interaction between students. The goal of the learner is to regurgitate the accepted explanation or methodology expostulated by the teacher.

Airasian and Walsh (1997) refer to the widespread tendency to set up straw men:

> First, the rhetoric that surrounds constructivism is seductive. It plays off the metaphor of "lighting the flame" of student motivation (constructivism) against that of "filling the bucket" of students' heads with facts (present methods). Constructivists claim that they emphasize autonomy as opposed to obedience, construction as opposed to instruction, and interest as opposed to reinforcement. The implication is that, if one is opposed to constructivism, one is opposed to student autonomy, construction of meaning, and interest. Thus, opponents are viewed as being against lighting the flame of student motivation. Such rhetoric plays a potent role in the reception of all innovations, including constructivism. (p. 447)

Is It Possible to Define Constructivism?

Of course, it is technically possible to define constructivism, but the question is whether it is possible to do so in a way that would produce any consensus among educators. We think it is probably not possible to arrive at substantial agreement about definition at the present time. It seems to us that a logical, coherent definition would be possible only by identifying the theorists whose work is basic to constructivism by carefully studying that work, and then identifying core concepts that are common to all. For example, if we agreed that Piaget and Vygotsky are the basic constructivist theorists, we would probably produce a definition that stressed the importance of sequential cognitive development, concept formation, interpersonal experiences, differences between concrete and abstract thought, the interaction between innate characteristics and the environment, etc. Once this was done, we might be able to (a) design educational assessments to measure children's cognitive developmental status, (b) develop teaching methods that would be developmentally appropriate, and (c) choose research problems and strategies that would help solve educational questions and problems.

However, we doubt this is a practical goal at this time. *Constructivism* has simply come to mean too many different things to too many people. The situation seems similar to that in special education with regard to the term *dyslexia*, which has come to mean everything from a neurologically based inability of middle elementary age children to read even simple words after competent and extended remediation, to very mild reading comprehension problems of adults. Because of this situation, it is impossible to know what kind of problem is being referred to when the term *dyslexia* is used. This lack of an agreed-upon definition of dyslexia even prompted one professional organization to recommend that the term be abandoned.

The same confusion prevails with respect to constructivism. The only agreement about what it is seems to be is that it involves children constructing their own knowledge. But this idea is so vague and general that it has no real practical value for educators or researchers.

The term is sometimes used to justify anything and everything. One example of this occurred at a recent international conference for information technology in education. One presenter cited the fact that he is a constructivist as the justification for suggesting that research is anything that the researcher says is research; abandoning quantitative research in favor of qualitative research; defining research as anything that promotes the emancipation of oppressed groups; rejecting the idea of objective reality or universal truths; and accepting poems, paintings, novels, short stories, or sculptures as educational dissertations.

Some or all of these ideas may have potential. However, we do not believe that they are closely related to the work of Vygotsky or Piaget, for example, and are not, therefore, constructivist in nature (if we accept Vygotsky and Piaget as constructivist theorists). Although both conducted research that might be considered qualitative in nature, they also designed and carried out studies that were quantitative, as well. Certainly, neither theorist could be legitimately characterized as taking a loose, or less-than-rigorous view of the nature of research, qualitative or quantitative. We have to wonder what Vygotsky or Piaget would have thought about the notion that science cannot reveal what is really going on in a learning situation, that research is whatever the researcher says it is, or that an educational dissertation should be anything that promotes the emancipation of an oppressed group. We would like to know which other theorists sometimes associated with constructivism would subscribe to all or most of these views.

Because of the multiplicity of meanings attributed to the term *constructivism*, we would suggest that the term be abandoned altogether. Because it means all things to all people, it actually loses its meaning entirely. Kauffman (1999) has suggested that the best paradigm for research is the one that "offers the most practical and reliable tool for problem solving" (p. 249). Constructivism, however, because it is so imprecisely defined, offers little in the way of a problem-solving tool and nothing in the way of a guiding light for research.

SUMMARY

Recently, there have been many calls for a new focus in education to be used to redefine schooling and research. This new focus ("paradigm," "theory," "philosophy," "method," etc.) has been termed *constructivism*, and has been used to justify a variety of sometimes-contradictory assertions and recommendations about education, teaching, and learning. Virtually the only agreement that can be found across even a small sample of books and articles about constructivism is the general notion that children construct their own knowledge. Otherwise, little consensus exists in the modern literature about constructivism.

To arrive at a logical definition, it seems reasonable to review the ideas of those theorists who are often cited as the founders of constructivism such as Piaget and Vygotsky. Such a logical definition might make the term more useful for designing curricula, devising teaching methods, or conducting and interpreting educational research. However, because few writers read and study the work of Piaget, Vygotsky, or other so-called constructivists, the term *constructivism* continues to be loosely defined (or, more often, undefined). It is therefore impossible to know what different writers are referring to when they use the term. We suggest that the term be abandoned, and that writers of position papers and research reports be specific about their philosophical and theoretical assumptions, rather than making reference to a term that is essentially meaningless.

REFERENCES

Airasian, P.W., & Walsh, M.E. (1997). Constructivist cautions. *Phi Delta Kappan, 78*, 444-450.

Bodrova, E., & Leong, D.J. (1996). *Tools of the mind: The Vygotskian approach to early childhood education*. Columbus, OH: Merrill.

Clements, D.H. (1997). (Mis?)constructing constructivism. *Teaching Children Mathematics, 4(4)*, 198-200.

Eisner, E.W. (1991). *The enlightened eye: Qualitative inquiry and the enhancement of educational practice*. New York: Macmillan.

Eisner, E.W., & Peshkin, A. (Eds.). (1990). *Qualitative inquiry in education: The continuing debate*. New York: Teachers College Press.

Hanley, S. (1994). On constructivism. [On-line]. Available: *http://www.inform.umd.edu:8080/UMS+State/UMD-Projects/MCTP/Essays/Constructivism.txt)*. College Park, MD: Maryland Collaborative for Teacher Education.

Hart, D., Kohlbert, L., & Wertsch, J.V. (1987). The developmental social-self theories of James Mark Baldwin, George Herbert Mead, and Lev Semenovich Vygotsky. In L. Kohlbert (Ed.), *Child psychology and childhood education: A cognitive-developmental view*. New York: Longman.

Jaramillo, J.A. (1996). Vygotsky's sociocultural theory and contributions to the development of constructivist curricula. *Education, 117(1)*, 133-141.

Kauffman, J.M. (1999). Commentary: Today's special education and its messages for tomorrow. *The Journal of Special Education, 32*, 244-254.

Kindsvatter, R., Wilen, W., & Ishler, M. (1996). *Dynamics of effective teaching* (3rd ed.). White Plains, NY: Longman.

Kohlberg, L., & Wertsch, J.V. (1987). Language and the development of thought. In L. Kohlberg (Ed.), *Child psychology and childhood education: A cognitive-developmental view*. New York: Longman.

Kozulin, A. (1990). *Vygotsky's psychology: A bibliography of ideas*. Cambridge, MA: Harvard University Press.

Lancy, D.F. (1993). *Qualitative research in education: An introduction to the major traditions*. New York: Longman.

Murphy, E. (1997). Constructivism: From philosophy to practice. [On-line]. Available: *http://www.stemnet.nf.ca/~elmurphy/emurphy/cle.html*

Ormrod, J.E. (1998). *Educational psychology: Developing learners*. Upper Saddle River, NJ: Prentice-Hall.

Piaget, J. (1923). *The language and thought of the child*. Atlantic Highlands, NJ: Humanities Press.

Piaget, J. (1924). *Judgment and reasoning in the child*. Atlantic Highlands, NJ: Humanities Press.

Piaget, J. (1995). The child's conception of physical causality. In H.E. Gruber & J.J. Voneche (Eds.), *The essential Piaget: An interpretive reference and guide*. Northvale, NJ: Jason Aronson.

Slavin, R.E. (1997). *Educational psychology* (5th ed.). Boston, MA: Allyn and Bacon.

Van der Veer, R., & Valsiner, J. (1991). *Understanding Vygotsky: A quest for synthesis*. Oxford: Blackwell.

Vygotsky, L. (1986). *Thought and language*. Cambridge, MA: The MIT Press.

Jerry Willis
Muktha Jost

Computers and Qualitative Research

Qualitative research is a very broad term. It can involve the collection of many different forms of data–from field notes to reflective journals to newspaper articles, diaries, and letters. Qualitative research can also involve an amazing variety of data analysis techniques. There is, currently, considerable debate on the proper ways to analyze qualitative data, and a sub-issue in this debate is the role computers can play in the data collection and analysis process. In this paper, we will first look at some relatively "uncontroversial" ways computers can be used in qualitative research. Then we will look at examples of qualitative research that reflect two frameworks for data analysis and interpretation. The two examples illustrate *holistic* and *atomistic* approaches. The implications of these two traditions in qualitative research, both in terms of how we approach data analysis and the ways computers can be used, will be a major focus.

NONCONTROVERSIAL USES OF COMPUTERS IN QUALITATIVE RESEARCH

Most of the controversy over computers in qualitative research centers on the use of computers to analyze qualitative data–interviews,

JERRY WILLIS is Professor, Curriculum and Technology, Center for Teaching, Learning, and Technology, College of Education, Iowa State University, Ames, IA 50010. E-mail: jerryw@iastate.edu
MUKTHA JOST is a PhD student, Curriculum and Technology, Center for Teaching, Learning, and Technology, College of Education, Iowa State University, Ames, IA 50010. E-mail: Mjost@iastate.edu

[Haworth co-indexing entry note]: "Computers and Qualitative Research." Willis, Jerry, and Muktha Jost. Co-published simultaneously in *Computers in the Schools* (The Haworth Press, Inc.) Vol. 15, No. 3/4, 1999, pp. 21-52; and: *Information Technology in Educational Research and Statistics* (ed: Leping Liu, D. LaMont Johnson, and Cleborne D. Maddux) The Haworth Press, Inc., 1999, pp. 21-52. Single or multiple copies of this article are available for a fee from The Haworth Document Delivery Service [1-800-342-9678, 9:00 a.m. - 5:00 p.m. (EST). E-mail address: getinfo@haworthpressinc.com].

reflections, notes, and historical documents to name but a few possible types of data researchers collect. There are, however, many other ways computers can support qualitative research. We have divided those other ways into three categories: sources of information and collaboration, sources of data, and ways of communicating results. A fourth use of computers, for data analysis, will be addressed in more detail later in this paper.

SOURCES OF INFORMATION/COLLABORATION

Two nearly universal aspects of scholarship are the search for relevant information (the "literature review") and efforts to collaborate and share with others that have similar interests. Both can be facilitated and enhanced by computer technology. Many researchers use standard electronic mail services to communicate with other scholars with similar interests, and they use E-mail to send documents to colleagues, editors, and individuals interested in their work. We will not discuss E-mail in detail here because most people are familiar with it. It is the simplest form of what has become known as *computer-mediated communication*, or CMC.

Electronic Discussion Groups

Another form of CMC that may be less familiar to some readers is electronic discussion groups. These groups, sometimes called *lists* or *news groups*, are ways people interested in a particular topic can "meet" electronically and discuss mutually interesting topics. There are hundreds of thousands of discussion groups on the World Wide Web today. A growing number of these discussion groups are of interest to qualitative researchers. Below is an annotated listing of a few examples:

1. *Biographical Methods Discussion List.* This discussion is for social scientists interested in the use of biographical research methods based on the interpretive paradigm of research. To join this discussion, send an E-mail message to *mailbase@mailbase.ac.uk* In the body of the message send this message, substituting your first and last names for the general terms in the message: *join biog-methods yourfirstname yourlastname*

If Kristen Wright were to sign-up for this list she would send the message *join biog-methods kristen wright*, and she would begin receiving messages from the group.

2. *ETHNO HOTLINE*. This discussion list emphasizes ethnographic research and frequently carries lively discussions of current issues. To subscribe, send an E-mail message to this address: *COMSERVE@CIOS.ORG* Enter this message, mail *JOIN ETHNO YOURFIRSTNAME YOURLASTNAME*.
3. *Qualitative Research for the Human Sciences*. This is a general qualitative research list that supports ongoing conversations about current issues and needs. To subscribe, send a message to *listserv@uga.cc.uga.edu*. In the body of the message type the following: *subscribe qualrs-l yourfirstname yourlastname*

If you are interested in one of the groups for qualitative researchers, the information provided above may be enough to get you started. Please keep in mind, however, that there are no universal rules for how E-mail is created, sent, and read. If you are not familiar with the E-mail system you will be using, it may be necessary to get some training and support from an experienced E-mail user before you can successfully join a list.

Once you have joined, you should begin to receive messages from the list. The first time you read messages, you may be tempted to jump right into the conversation. There is an etiquette to participating in a list, and we suggest you *lurk* (read what other people are saying but do not reply yourself) until you feel sure you have a grasp of the rules of the list as well as appropriate topics for discussion.

The three lists described above are a tiny sample of the several hundred discussion groups available for qualitative researchers. We will not attempt to list or describe them here, in part because they change so quickly. It is possible, for example, that by the time you read this, none of the lists mentioned earlier will be in operation. However, there is a way to get current information about active discussion groups. Several qualitative research Web sites keep relatively up-to-date information on discussion lists.

Qualitative Research Web Sites

In March of 1999, we used a popular *search engine* (which will be explained in the next section), AltaVista, to look for World Wide Web sites that contained the term *qualitative research.* There were 18,280 "hits." Each hit represented a Web site containing the term *qualitative research.* The World Wide Web (WWW or Web) has become a rich information resource for qualitative researchers. The Web is so rich, in fact, that the sheer volume can be overwhelming. You can deal with that problem in several ways. One approach is to narrow your search. When we searched for sites that contained the following three terms:

> "qualitative research,
>
> interpretive, and
>
> education",

for example, there were only 2,952 hits. When we added *dropout* to this list of search terms there were only 17 hits.

Narrowing your search criteria is a very good way to locate Web-based content on a specialized topic. But what if you have a general interest in exploring qualitative research but don't want to wander through the 18,280 hits a search produced? There is, fortunately, a simple solution–*gateway Web sites.* There are several Web sites about qualitative research that contain many different types of information as well as annotated lists of other useful and interesting Web sites. The best of these is QUALPAGE, which is maintained by a nursing faculty member, Dr. Judy Norris, at the University of Alberta. QUALPAGE is divided into 15 different categories:

1. Announcements: Calls for abstracts, conferences, workshops
2. Approaches
3. Conceptual and philosophical foundations
4. Disciplines
5. Discussion forums
6. Electronic journals
7. Grants information
8. Methods
9. Organizations and interest groups for qualitative researchers
10. Other resources for qualitative researchers
11. Papers
12. Conference proceedings
13. Presentations of research on the Internet

14. Publishers of journals and books for qualitative researchers
15. Qualitative Data Analysis (QDA) software resources.

QUALPAGE contains links to everything from introductory material on relevant theories and research methods to conference papers that report the latest research. It is probably the most useful and best known general Web resource for qualitative researchers. The electronic address (or URL, universal resource locator) for QUALPAGE is: *http://www.ualberta.ca/~jrnorris/qual.html*

Another gateway Web site is the Qualitative Research Page, which is maintained by Dr. Bobbi Kerlin. The address is *http://www.ualberta.ca/~jrnorris/qual.html*. This site is organized around 13 types of information:

1. Research discussion groups
2. Course syllabi for qualitative research courses
3. What is research?
4. Research ethics
5. Theory and its role in research
6. Research paradigms | genres | methods
7. Information gathering: Includes general references, conducting research on the Web, Canadian resources, and ERIC research resources.
8. Planning and preparing research
9. Interpreting and representing research
10. Research software
11. Theses and dissertations on the Web
12. Links to other qualitative research resources
13. The quant corner

A third Web site, run by Ron Chenail, a professor in the School of Social and Systemic Studies at Nova Southeastern University, is another very valuable resource. The Web address of the Qualitative Inquiry Project is *http://www.nova.edu/ssss/qualinq.html*. At that location you will find two different resources. *The Qualitative Report* (*http://www.nova.edu/ssss/QR/index.html*) is an electronic journal that is available to everyone via the Internet. Chenail also supports a Web site called Qualitative Research Resources on the Internet (*http://www.nova.edu/ssss/qualinq.html*). This site contains information on:

1. Other Web pages of interest to qualitative researchers
2. Papers and other textual sources
3. Calls for qualitative research papers, abstracts, and proposals
4. Ethical aspects of qualitative research in health research
5. Release announcement
6. Qualitative research syllabi

Chenail's site has one of the largest lists of discussion groups and Web sites relevant to qualitative research. Those range from *Action Research in Education* and *Action Research Resources* to *Theory Construction and Research Methods.*

For most of the resources listed on these gateway sites, the links are "hot." For example, if you click the listing for *The Participant Action Research Network* at Ron Chenail's Web site, you will be taken to the home page (the main or first page of a Web site) of the Participatory Action Research Network. Participatory action research is a form of action research based on interpretive and critical theories instead of empirical and positivist paradigms. This site contains introductory information on participatory action research as well as coverage of advanced topics, discussion groups, and links to other Web sites about action research and participatory inquiry.

Together, the three gateway sites introduced here provide a beginning qualitative researcher with connections to hundreds of resources that have been reviewed and recommended by people with expertise in the field. You may still want to search the Web for specialized resources, but we recommend you start with these because they have been selected from a large number of sites for their quality and usefulness. For those specialized searches, however, you will need to use a search engine.

Internet Search Engines

In the early years of the Internet, there were very few information resources on the Net. Those that were on the Internet tended to be specialized papers and reports that appealed to small, tight groups of scholars who shared an interest in one form of specialized research or another. Today, as the ad says, "We've come a long way." The problem with the Internet today is not that there isn't likely to be anything on a topic that interests you. The problem is that there are billions of documents on the Web and sometimes it is hard to find exactly what you're looking for because the other two billion documents keep getting in the way. This embarrassment of riches led to the creation of a special form of software–search engines. A search engine is a system that has two parts:

1. A specially constructed database of the content of documents available on the Internet. (Some search engines only have World Wide Web sites in the database; others include other types of Internet resources such as discussion groups.)
2. A user interface that lets you specify terms you want to look for on the Internet.

Search engines vary in the number of sites in the database and in the form and sophistication of the user interface. Three popular search engines are:

AltaVista	*-http://www.altavista.com*
Infoseek	*-http://www.infoseek.com*
Snap	*-http://www.snap.com*

There are many other search engines, but for most searches these engines should do the job. They all work in approximately the same way. Go to the Web site, type a word, phrase, or set of words, and click the Search button. The engine will then return a list of "hits." Then you can click on the name or title of a hit and go to that Web site.

If you would like to explore the more than 100 search engines that are operating on the Web today, you can go to the Search Engine Watch Web site (*http://searchenginewatch.com/*) and read comparative reviews as well as use links to go to search engine sites. Most browsers (like Netscape Communicator and Internet Explorer) also have a button you can click that will take you to a list of search engines.

Most sophisticated search engines support complex searches that make use of Boolean logic and special characters to help you formulate a very precise search. Below are some examples that work with Infoseek, but other search engines have similar features.

+social_class +employment. This search has two terms in it. Putting a + in front of each term means a hit must have that term in it.

Because both terms have a +, this search will only report hits that contain both the term *social class* and the word *employment*. The underline character (_) between social and class is a connector. It puts the two words together to indicate you want social class treated as one term instead of two different words. If you left the underline out and there was a space between social and class, the search engine would look for hits with the word *social* and the word *employment*, and it would put at the top of the list hits that also have the word *class*.

"International Business Machines." Without quotes this search would locate Web sites that have the three words in the search term. However, they would not necessarily be in the order listed in the search. Put quotes around the three terms and the search will select only sites that have the term *International Business Machines* in them. A site with *International* in one paragraph, *business* in another, and *machines* in yet another would not be considered a hit.

To summarize, computers can help develop and support collaborations between groups of researchers and scholars. At the lowest level, E-mail can be a powerful and convenient way of communicating. Other forms of computer-mediated communication, including lists and discussion groups, are also very useful. They can put scholars interested in similar topics in easy contact even if they live and work in different states, or countries. The World Wide Web has also become an excellent information resource for qualitative researchers. There are thousands of sites on the World Wide Web that have relevant information. There is, in fact, so much information that users need ways of simplifying the process of finding the type of information they need. There are a number of gateway Web sites for qualitative researchers that have organized and annotated the best qualitative research sites. You can use a gateway site to make decisions about what sites you want to check out. And, as you become interested in more specific topics, you can use one of the search engines to look for relevant Web sites yourself.

ELECTRONIC SOURCES OF DATA

A second, relatively noncontroversial use of computers is as a source of data. A great many sites on the Web have many different forms of quantitative data. For example, the Council on European Social Science Data Archives maintains a Web site (*http://www.nsd.uib.no/cessda/index.html*) that lists the major data archives for Europe and North America. This site, for example, contains links to the Finnish Social Science Data Archive (in Finnish) and the Harvard-MIT Data Center. The Harvard-MIT site provides access to a number of data archives including a number of ABC News/Washington Post polls conducted by the Roper Organization.

While most of the data archives available on the Web today house quantitative data such as survey responses scored on the familiar Strongly Disagree to Strongly Agree scale, a few sites now make qualitative data available. One of the most active sites for archiving qualitative data is Qualidata (*http://www.essex.ac.uk/qualidata/*) at the University of Essex in the United Kingdom. This site handles data from interviews, observations, and other forms of qualitative research. The site is supported and maintained by the Department of Sociology. You can submit data for archiving as well as gain access to data already linked to their system. Qualidata is relatively unique today, but there are likely to be many more resources like this on the Web in the near future.

DATA COLLECTION

Desktop computers are too bulky to take into most research environments, but they have been used for years to transcribe field notes and interview data during the process of data collection. A researcher might, for example, make handwritten notes throughout a day of observation and then transfer those notes (and add comments) that evening using a desktop computer with a word processor.

With the arrival of light, battery-operated lap top computers many researchers find it convenient to take their computers with them when conducting observations. Instead of writing their notes on a ruled pad they type them into a word processor on their laptop. This reduces the amount of work required to get the data into the computer file, but researchers should be careful not to be disruptive. In many settings, the clicking of keys is annoying and irritating while writing on a pad is not.

Finally, two emerging technologies offer considerable promise in the future for data collection. One is the computer pad. Cross's electronic writing pad is in the same shape as a standard writing pad, but, when you write on the surface of the pad, it converts your handwriting into text data that can be exported from the pad to a computer. These electronic writing pads are not quite powerful enough to be routinely used in qualitative data collection today. They make too many mistakes and their memories are not large enough to hold a full day's notes. They do show promise, however, and it may not be long before they are standard equipment for participant observers.

The other emerging technology is voice recognition. At least 10 companies now sell programs that let you speak to the computer (in true *Star Trek* style). The computer will obey spoken commands and enter dictation into a word-processing file. Again, these programs are not yet good enough to be used routinely, but within a few years it may be possible to conduct an interview and let one of these programs transcribe the spoken words directly to a word processor file. Again, however, current models are error prone and temperamental. They are more a sign of what is coming than useful products for qualitative research.

COMMUNICATING RESULTS

A third, relatively noncontroversial use of computer and information technology is to support efforts to communicate the results of qualitative research. Few researchers prepare manuscripts for submission to journals today without using a word processor. An increasing number of journals require manuscripts be submitted in electronic form because it saves the publisher time and money. Today, word processing has become as ubiquitous as an ink pen was 150 years ago. But that is only the beginning of ways computer technology can support communication.

Desktop Publishing Programs and the Publication of Journals

Microsoft Word and *WordPerfect* are familiar to thousands of authors who use these word-processing programs to prepare their manuscripts for publication. Authors may not realize that another type of software, desktop publishing, is used by the editors and publishers of thousands of journals, proceedings, monographs, and books today to prepare the pages that will become the printed document. Programs like *PageMaker*, *Ready-Set-Go*, *Ventura Publisher*, and *QuarkExpress* go well beyond what can be done with ordinary word processors. Desktop publishing programs reduce the cost and the time required to produce publications with a professional, polished appearance. For many years, for example, I was part of a team that produced the *Technology and Teacher Education Annual* each year using *PageMaker* to format the book (which grew to well over a thousand pages) between December and an annual conference the following March. Using equipment and software that costs less than $2,500, small organizations, centers, individuals, and publishers can produce excellent publications. The publisher uses the desktop publishing program to create all the pages in the publication, prints them on a high-resolution laser printer, and then contracts with a printer to produce the document in quantity.

Hypermedia and Multimedia Documents

Traditionally, most qualitative research has been disseminated primarily as linear text, with the occasional figure or illustration. By linear, we mean text that is generally intended to be read from the beginning to the end in a linear sequence–one section or chapter after another. Some studies, and some types of data analysis, do not lend themselves to linear text. Some reports of research would profit from the ability to include audio, video, or animated graphics. Multimedia development environments such as *Authorware*, *Director*, and *Hyperstudio*, to name just three, require some effort to learn to use. Once mastered you can create documents that have much more than text. A discussion of an interview, for example, could be enhanced by an audio or video clip from the interview. A discussion of the differences in the condition of inner city and suburban school buildings and the impact that has on learning can be done with text and a few pictures, but a *virtual reality* walk through some of the schools will have much more impact. Qualitative researchers may increasingly use multimedia papers and monographs in the future to communicate with readers. The journal that publishes these papers may be distributed on a CD-ROM or over the Internet. (Many authors today also put drafts or "preprints" of their scholarly papers on their own Web sites before submitting them for publication. Anyone visiting those Web sites can select papers from a list and then read them online, print them on their own printer, or download them to be viewed or read later.)

Adding multimedia to a text-only research paper can be an improvement. However, the resulting multimedia paper is still usually linear. You begin at the introduction and read through to the conclusions. Another alternative is to make the paper *nonlinear hypertext.* A hypertext paper (which may have multimedia elements as well as standard text) contains links that let you move from one section of the paper to another. Hypertext papers can be read from beginning to end, but they are usually designed to be read in nonlinear fashion, with each reader following links that interest him or her. Although there have been a number of specialized programs for creating multimedia hypertext, few, if any, have been financially successful. The easiest way today to create hypertext is to use one of the programs for creating World Wide Web sites. Programs like *PageMill*, *FrontPage*, and *Claris-Page* let you create multimedia hypertext documents that can be placed on the Web or distributed on disk or CD-ROM.

Electronic Publication Options

Printed journals and monographs are still the primary means of communicating qualitative research reports, but this ink-on-paper approach faces growing competition from electronic publications that are available on the Web. Three such electronic journals are: *The Qualitative Report* (*http://www.nova.edu/ssss/QR/index.html*), edited by Ron Chenail at Nova Southeastern University; *Sociological Research Online* (*http://www.socresonline.org.uk/socresonline/*), edited by Elizabeth Stanley at the University of Manchester in the United Kingdom; and *Social Research Update* (*http://www.soc.surrey.ac.uk/sru/Sru.html*), edited by Nigel Gilbert, from the Department of Sociology at the University of Surrey in the United Kingdom.

All three of these journals are available to anyone who has access to the World Wide Web, and they frequently publish papers on issues related to qualitative research methodology. These and other electronic journals (e-journals) are becoming significant outlets for a wide range of qualitative research. And, in general, the e-journals tend to be more open to innovative types of research and ways of reporting the results.

While the four journals mentioned earlier are all focused on qualitative research, there are hundreds, perhaps thousands, of other e-journal, newsletter, and occasional paper publishers that use the Internet to distribute accepted manuscripts. The gateway Web sites mentioned earlier are a good source of information for e-journals about qualitative research. For a list of over 7,000 electronic scholarly resources, go to the home page of the *Directory of Electronic Journals, Newsletters and Academic Discussion Lists* (*http://www.arl.org/scomm/edir/*). This directory, which is published by the Association of Research Libraries, is the most comprehensive source of information on electronic publications. It is available in print and electronic form, but access to the directory over the Web is by subscription only. However, a university that subscribes can provide access to all students and faculty. An excellent free Web site that lists many e-journals as well as newsletters is New Jour (*http://gort.ucsd.edu/newjour/*), which is housed at the University of California, San Diego. New Jour maintains a list of existing e-journals and regularly adds new publications to the list. This service began in 1993 with about 250 sources and today it has more than 5,000. This site is managed by Ann Okerson, Yale University, and James O'Donnell, University of Pennsylvania. It is one of the friendliest and easiest to use of the e-journal sites. This site has links to everything from the *American Journal of Psychiatry* to *Earthworm Growers of the World.* Most listings are *hot*, which means you can click the name of a journal and jump to the journal's home page.

Another free site is the BUBL Information Service (*http://bubl.ac.uk/*) from the Andersonian Library at Strachclyde University in Scotland. This site includes a list of several hundred journals, organized by content area. The abstracts or full text of over 200 journals are available on this site.

Finally, the Electronic Journal Alliance (*http://www.coalliance.org/ejournal/*) provides information on journals as disparate as *Abdominal Imaging* and *Accelerator Machine Physics* to the *Yale Human Rights and Development Law Journal* and the *Zoological Journal of the Linnean Society.*

As the cost of creating printed journals continues to increase, and the cost of subscribing to them increases at an even faster rate, more and more scholars will consider making their work available to a wider audience by publishing in one of the growing number of e-journals. And, as small groups and organizations find that papers from their group are not being published in established journals, one option is always to start another journal. In the past this has often meant raising substantial sums of money to start a small publishing company, produce the journal, and build a subscriber base. Today, because several large and medium-size companies have come to dominate scholarly publishing in many fields, starting a new paper journal increasingly often means finding a commercial journal publisher willing to take on the considerable cost of launching a new print journal. As journal publishing has become an increasingly commercial enterprise, subscription costs have soared. An inexpensive alternative is the e-journal. Today small groups, even individuals, can launch an e-journal using computer resources that are readily available in many faculty offices. Most universities offer space for Web sites on their central computers, so the only equipment needed to start a new e-journal is a personal computer with software for creating Web documents and a connection to the Internet. A system costing less than $800 could be used if you were operating on a shoestring.

Starting your own journal may not be as far-fetched as it sounds. Each year hundreds of journals cease publication because they no longer appeal to their audiences, and each year hundreds of new journals are launched. Over a period of 25 years I have been involved in the creation of four different journals, and all are still publishing. Launching a journal naturally involves much more than assembling the right computer system. It requires skills to rally interest, develop a mission or vision, get commitments from associate and contributing editors, and build a strong board of reviewers. It is hard work and takes considerable time, but it can also be very rewarding. Most journals are launched because the founders felt there was not a suitable publication outlet for a particular type of research. B. F. Skinner, for example, helped found the *Journal for the Experimental Analysis of Behavior* when mainstream psychology journals would not publish his studies of pigeons pecking metal disks to get pellets of food. And, the journal *Qualitative Inquiry* was recently launched because the founders felt there was a need for an outlet for papers on the methodology and issues of qualitative research.

Starting a journal is not the first thing you will want to do as an emerging scholar, but there may be times when it is an appropriate focus, and computer technology can make the task easier and less expensive.

QUALITATIVE DATA ANALYSIS AND COMPUTERS: AN ONGOING DEBATE

The use of computer resources to locate relevant information, develop collaborations, and communicate results of research is not very controversial today. That cannot be said for another way of using computers in qualitative research–data analysis. There is no shortage of opinions about the role of computers in qualitative data analysis. A number of authors have treated the arrival of computerized systems of qualitative data analysis as a salvation. They argue that these programs make possible more sophisticated analyses while at the same time making the process easier and less frustrating.

On the other hand, some have responded to the increasing use of computers in qualitative data analysis quite negatively (Coffey, Holbrook, & Atkinson, 1996). The creation of qualitative data analysis software has even been compared to the creation of the Frankenstein monster which, once created, turns against its makers (Kelle, 1997; Lee & Fielding, 1991). A reading of this literature suggests there are three major reasons for the diversity of opinions about the usefulness and value of computers in qualitative data analysis.

Past Associations. For some, the issue is one of past associations. Qualitative researchers tend to take the side of the underdog, the oppressed, and the underrepresented. And, they tend to adopt interpretive and critical worldviews as opposed to modernist, empiricist, and positivist perspectives (Smith, 1993). For much of the history of computers, since they were developed in the mid-1940s, computers have been associated with big business, big government, and the Establishment which championed modernist and empiricist world views. They have also been championed by positivist and postpositivist researchers as powerful tools for advancing the very types of research that interpretive and critical researchers sometimes find less than appealing. Thus, when computer programs purporting to support "qualitative" research arrive on the scene, it is not surprising that they are greeted with some degree of skepticism.

In our opinion, this problem is with the way computers are sometimes used, not with all ways they can be used in research. The uses of computers anyone considers "appropriate" and "inappropriate" will depend on the paradigm they use. Postpositivists, for example, will analyze their data in ways that are quite different from the procedures used by qualitative researchers in the interpretive and critical traditions. Thus, their view of which programs are "good" and which are not, will vary.

Early Numerical Focus. Another reason many qualitative researchers find the use of computers unappealing is the strong historical association between quantification and computers (Kelle 1997). Computers have a long history of number crunching, and all the well-established programs–such as SPSS, SAS, and BMD–for analyzing data were created specifically to analyze quantitative data–numbers. Qualitative researchers, who have had to struggle for decades to have their work accepted in the mainstream of social science and education research, may feel that using computers is selling out–adopting the very quantification model commonly associated with computers that they were trying to move beyond. And, because it is easy to do that type of analysis with a computer, they see the computer as an accomplice in pushing researchers in that direction. As Coffey, Holbrook, and Atkinson (1996), put it,

> The categorization of textual data and the use of computer software to search . . . [for patterns in the categories] appear to render the general approach akin to standardized surveys or experimental design procedures. In our view qualitative research is not enhanced by poor imitations of other research styles and traditions. Analytic procedures which appear rooted in standardized, often mechanistic procedures are no substitute for genuinely 'grounded' engagement with the data throughout the whole of the research. (Online)

There is a point to this concern. Some qualitative researchers have adopted a perspective that is very close to that of traditional quantitative researchers (Miles & Huberman, 1994a, b), and they tend to be much more comfortable with the idea of carrying out quantitative analyses that involve statistical tests of numerical data derived from qualitative sources such as interviews or observation notes. Many software programs are available today that support this type of work, and the ease with which they can be used to analyze data is a temptation, even for researchers who are trying to work within alternative paradigms.

Links Between Grounded Theory and Computer Analysis. One of the most interesting and well-considered critiques of qualitative computer analysis is a paper by faculty at the School of Social and Administrative Studies at the University of Wales (Coffey, Holbrook, & Atkinson, 1996). Their argument is relatively straightforward. They say there has been a "postmodern turn" in qualitative research that has led to the questioning of established methods of data analysis and the exploration of alternatives. Coffey et al. suggest that a number of movements are contributing to a trend in qualitative research toward multiple sources of data,

- multiple ways of both analyzing and thinking about data,
- multiple ways of communicating the results of qualitative research, representing multiple voices and perspectives,
- and mixed, complex ways of thinking about social issues.

The purpose of the type of research Coffey and her colleagues support is not to narrow what we can say to one "true" account. Instead, it is to share the rich diversity of viewpoints, perspectives, and opinions. They feel that in the future, hypertext software may be a solid foundation for this type of data analysis as well as a means of communicating the results of research.

Coffey and her colleagues believe the general trend toward diversity is in marked contrast to the influence of computers in qualitative research. They believe the arrival of sophisticated computer programs for qualitative analysis has had the unfortunate effect of pushing us toward homogeneity instead of diversity. "An emerging orthodoxy is being adopted globally by key members of the qualitative research community. This is largely, though by no means exclusively, linked to the growth of Computer Assisted Qualitative Data Analysis Software (CAQDAS) as a subfield of expertise. . . ." (Online). These authors believe most CAQDAS supports one general model of analyzing data–coding. "The computer-based approaches of this sort depend on procedures for coding the text (interview transcripts, fieldnotes, transcribed recordings, documents). This means marking the text in order to tag particular chunks or segments of that text. Codewords are thus attached to discrete stretches of data" (Online). Coffey et al. use the Ethnograph as a prototypical example of software that helps researchers analyze their data by coding. It, and other programs, are technical advances over manual methods because coding is faster, and searches to retrieve instances of a particular code or pattern of codes are thorough. In addition, the software "can cope with multiple and overlapping codes; it can conduct multiple searches, using more than one codeword simultaneously. Software like the Ethnograph allows the analyst to combine codewords, in an approximation of Boolean logic, to facilitate complex searches" (Online).

Coffey and her associates further note that qualitative software analysis packages like NUD*IST, the Ethnograph, and Kwalitan also allow the researcher to attach memos to any part of the data. These memos may be anything from questions to hypotheses and tentative theories that might identify and explain patterns in the data.

These two features–support of sophisticated coding and retrieval and the ability to add memos to the data–are fine features of CAQDAS, but Coffey et al. see these as characteristics of one approach to qualitative research–*grounded theory.*

> Aspects of grounded theory have been over-emphasized in the development and use of qualitative data analysis software, while other approaches have been neglected in comparison. There is, therefore, a danger that researchers may be led implicitly towards the uncritical adoption of a particular set of strategies as a consequence of adopting computer-aided analysis. The emphasis on coding data is a central feature of this process of convergence (Online).

In essence, the grounded-theory approach involves collecting and analyzing data, developing a tentative theory or hypothesis to explain the data, collecting more data to test the theory, and then using the new data to revise and improve the emerging theory. Mulder's (1994) work, which is described in detail later, used this approach and several qualitative software packages, including the Ethnograph and NUD*IST, support this approach.

Coffey's contention that the use of CAQDAS pushes us toward the use of a single approach to analysis–grounded theory–is one of many expressions of concern about the unsavory influence of computers in qualitative research (see, for example, chapters in a book edited by Lee & Fielding 1991). It has also been strongly disputed by others. Kelle (1997), for example, points to differences in paradigms as the source of the concern about whether computers should be used in qualitative data analysis.

> A closer look at the philosophical and epistemological roots of interpretive research makes clear that a certain caution against computer technology is justified with regard to the nature of the process of hermeneutic verstehen. Philosophical approaches which play an important role within qualitative research, such as Phenomenology, The Oxford Philosophy of Language and continental Hermeneutical Philosophy . . . had always stressed that ambiguity and context-relatedness have to be regarded as central characteristics of everyday language use. Following this argument–which has been further elaborated by contemporary post-modernist approaches . . .–it is impossible to make sense of written or spoken messages in everyday contexts–an operation which forms the core of hermeneutic Verstehen–without a 'tacit knowledge' which cannot easily (if at all!) be formalized (Kelle, 1997, Online).

Kelle compares this approach, which we will call "holistic" to an alternative that involves "the formulation of exact and precisely stated rules, which are completely context-free and contain no ambiguities" (Online). These two frames–holistic analysis that moves us toward contextual understanding versus analysis that breaks down the data and strives to develop rules and laws that are context independent–represent two major approaches to qualitative analysis. These alternatives, for example, are represented in the two cases (Collins, 1998, and Mulder, 1994) discussed in the next section of this paper.

While code and retrieve is the dominant way qualitative data is analyzed today, there is another approach that helps researchers relate or connect different parts of their data. The other approach, *electronic cross referencing*, relies on hyperlinks that let you jump from one area of the data to another. "By pressing a 'button' the user of a textual database can jump between the text passages which are linked together" (Kelle, 1997, Online). Kelle points out that, while the concept of hyperlinks seems new, they have been in use for hundreds of years. "One can easily see this by opening an ordinary King James Bible where a multitude of 'hyperlinks' are displayed on the margins of every page. By using these links a 'bible user' can, for example, jump between a teaching of Jesus in one of the Gospels and a passage of the Old Testament to which Jesus refers in this teaching" (Online). This form of data analysis (synopsis), which involves looking at sets of related text segments located at different places in your data, has generated important insights in many fields (Kelle, 1997). One qualitative program, ATLAS/ti, is particularly useful for creating several different types of links, including chains, loops, and complex networks of connections or links that are best represented graphically. And, because ATLAS/ti supports the creation of notes that are added to the data by the researcher, it also allows you to include those notes or memos in your pattern of hyperlinks.

To summarize, the way computers are used in qualitative data analysis depends on the particular framework the researcher adopts. This is a critically important point and to make it even more strongly, we have included two case studies that illustrate holistic and atomistic methods of analyzing qualitative data.

Case 1. "Unstructured" Interviewing

Peter Collins (1998) was concerned with the experiences of adults encountering chronic job insecurity. His research involved a series of open or unstructured interviews with individuals who worked for a local government in the United Kingdom that was undergoing a major restructuring. The changes dislocated many workers. Some were transferred to other agencies, some lost their jobs, and many were caught for months in the bureaucratic wasteland between the announcement that changes were coming and the arrival of those changes. Collins, who teaches qualitative research methods, points out that many students see interviews as "smash and grab" opportunities "in which they accost some innocent bystander and relieve them of whatever useful 'data' they may have" (Online). The purpose of interviews in this framework is for an expert to locate and mine factual information that can then be reported to other experts. Collins argues that this view of interviewing is encouraged by much of the literature on how to conduct interviews. It is, he believes, based on a number of assumptions including a structuralist model of language that sees language "as external, determined and determining, rather than language as an emergent social activity" (Online).

> The interview serves to facilitate the process wherein messages are conveyed from one mind to another. The interviewer encodes a thought in words and the interviewee decodes them to discover the thought and then similarly transmits a thought in turn. . . .

> Language is reduced to little more than a medium or transparent system which conveys messages from one mind to another. The structuralist model is essentialist, ignoring the way in which people actually talk and obscures the processes by which language is socially constituted. It perpetuates the myth that interviews are, above all else, opportunities for collecting objective facts. (Collins, 1998, Online).

Collins sees interviewing as a process of creating and constructing data rather than simply collecting it. To him, the interviews

> Are social interactions in which meaning is necessarily negotiated between a number of selves (and in which power may be *more* or *less* shared). The interviewer need be neither 'objective' nor 'detached,' but should rather be 'engaged.' Engagement implies a willingness on the part of the interviewer to understand the interviewee's response to a question or prompt in the wider context of the interview(s) as a whole. (Online)

For Collins, the interviews he conducts are not fact-finding missions, they are negotiations between individuals who are creating understanding through the conversation.

> My interviews comprise accounts of events together with attempts to interpret them on the part of interviewer and interviewee, but the process is haphazard and tentative. Rather than facts (which have an existence independent of the means of their discovery), such exchanges precipitate narrative: narrative that is emergent and indexical. Events and experiences are constituted, partly at least, in their telling (and re-telling). To describe a situation is to constitute it (as ethnomethodologists might put it). (Online)

Collins uses the term *indexical* to describe discourse in which terms and phrases used at one point are representations of or references back to other parts of the discussion. The narrative story he is building is *emergent* because it does not spring forth fully formed after Collins has gathered his data. Instead it gradually emerges as he reads and rereads his data and thinks about the data.

What does Collins publish when he finishes his research? It cannot be facts, since they are not available in any absolute sense. What Collins publishes is *narrative*, or "life stories." But if these stories are not truth, what are they? Collins quotes Chase (1996): "The aim of narrative analysis is not to impose immutable or definitive interpretations on participants' stories or even to challenge the meanings participants attach to their stories. Rather, its goal is to turn our attention elsewhere, to taken-for-granted cultural processes embedded in the everyday practices of storytelling" (p. 55).

Chase's emphasis on seeing the givens or assumptions in a story, and pointing them out, is one way of adding value to the story told by an interviewee. However, Collins points out that this is not the only way the researcher influences the story that is told.

> In encouraging the interviewee to tell me these stories and in asking them to develop a sub-plot here and a character there, I am encouraging them to construct and reconstruct themselves and contribute to this by exchanging stories of my own. . . . As I take less seriously the manuals' advice to maintain a lofty silence, I am increasingly moved to contribute my own stories, to hold them up for contrast or comparison to those of the interviewee. (Online)

Collins does not try to remain an objective outsider who maintains a degree of distance and objectivity. Instead, as the interviews progressed it "was increasingly difficult to remain passive, merely a listener. I was expected to participate, to contribute, to comment, to help in the development of stories which grow ever more intricate" (Online).

Collins' reports of the lives of adults experiencing job insecurity are several things:

1. They are subjective summaries of the stories told to him. Collins reduced many hours of interviews to write his reports. In doing so, he made many subjective choices about what to tell.
2. They are, in Chase's words, efforts to make more apparent and obvious the "taken-for-granted cultural processes" that were in the stories the interviewees told.
3. They are reflective analyses of the stories Collins and the interviewees have constructed together.

If we take Collins' approach to the analysis of interview data as a prototype, what sorts of data analysis did he perform and how could computers help in this type of work? Collins took a holistic approach–he did not break down the data into small bits that could be statistically analyzed. Instead, he combed the data over and over again looking for patterns and points. This holistic approach, which emphasizes developing meaning in context, is not facilitated by statistical software packages, but there are a number of ways qualitative software could help with the data analysis. To name just one way, programs like NUD*IST and ATLASt/i have a memo feature that lets you add any number of comments or notes to the raw data. As Collins' read and reread his data, the ability to add notes and comments to the data would be useful. The final section of this paper includes several other ways different types of computer software can support holistic qualitative research. In a companion paper, the use of computers to support atomistic qualitative research will be explored.

Case 2. Word Processing and Qualitative Data Analysis

In her master's thesis about consumers' opinions of family law at Curtin University in Australia, Mulder (1994) collected 37 interviews, which were entered into a data file that could be used by a qualitative research software package. Her goals were (a) to identify viewpoints expressed in the interviews that were shared across interviewees, and (b) to examine instances where different interviewees expressed opposing viewpoints on the same issue.

Mulder decided to analyze her interview data using a popular qualitative data analysis software package, the Ethnograph. On her first pass through her data, she coded the interviews for the occurrence of viewpoints on one of many issues that interested her. Comments on this one issue occurred in every interview but were not frequent. When Mulder ran a printout on this one issue, Ethnograph produced 74 pages! Mulder had two problems with the way Ethnograph handled her data. First, 74 pages of information on only one issue (from only half her interviews) was overkill. It actually complicated the data analysis task instead of simplifying it. Also, Mulder felt that "the program separated me from the words and experiences of the interviewees. I could not maintain an understanding of the context, compare incidents or consider emerging patterns while I was doing the analysis with the Ethnograph" (Online). Mulder used an emergent approach to her data analysis. As ideas came to her during the data analysis she reformulated her codes and the way she grouped the data, but that took many, many hours using Ethnograph. After several months of struggle, she came up with an alternative that worked well for her.

Mulder converted her data files to word processor documents and used Microsoft Word's outline view to create nine levels of headings to organize her data. Mulder used Ethnograph to create a word-processing file for each of the codes she was interested in (but this could have been done in Word had she began her analysis with Word instead of Ethnograph). That file contained every instance of that code (and the interview text associated with it). Then she wrote a short summary of each occurrence of that code and placed it at the beginning of the text segment. She made the summary headings using Word's outline view. When Mulder finished her initial pass, her data for each code looked like this:

Level 1 Heading	-Summary
Level 2 Heading	-Interview Number
Body Text	-Text associated with the code

Mulder then used Word's outline view to look only at the Summary Headings (the text was hidden). After some adjustments, she began to move the summaries around so that clusters of headings with similar meanings were together. (Moving headings also moved the associated text even though it was invisible.) Through this labeling and clustering process she became very familiar with the data. "By the time I had finished the labeling I knew the data very thoroughly, and as I moved similar themes together with continuous comparisons and criticisms, embryonic theories were being formulated" (Online). Mulder's final step was to create generalizations that explained these clusters. She refers to this as her emerging theory phase. "For example, when I discover an idea which explains three groups of data, I demote the levels of the groups and impose the new theory as a Level One heading. The structure of the developing analysis could be said to resemble a growing pyramid: the lower levels of headings contain original labels [summaries] and interview data, and the emergent analytical constructs are represented by increasingly higher levels of headings" (Online). Mulder used these constructs as the framework for writing her thesis.

Mulder's four stages of analysis–coding, writing summaries, grouping or clustering, and constructing theoretical interpretations–were all supported by computer software.

Data Analysis in the Collins and Mulder Studies

The Collins and Mulder studies are both clearly qualitative, but they represent quite different approaches to data analysis and to the ways computers can be used. While both studies are based on interviews, the purpose of conducting those interviews was different and the way the interview data were analyzed was also different. For Collins, the purpose was to provide readers with a narrative–a story. For Mulder, the purpose was to abstract out of the interview transcripts explanatory concepts–theories–that would help us understand in a much more general and universal way the feelings of consumers about family law.

Different purposes led to different approaches to data analysis. Collins took a holistic, contextual approach. Mulder also felt context was important, but she did break the interview text down and organize it according to a set of codes she developed. Her approach is a variation on the code-and-retrieve method of data analysis that is widely used in qualitative research today. The important point here is that there are many ways to approach qualitative data analysis and therefore many ways to use computers in qualitative research. The ways Mulder used computers would not be appropriate for Collins and vice versa. In the final section of this paper, we will look at a number of ways computers can support holistic qualitative research. (A companion paper looks at software that supports the atomistic tradition.)

SUPPORTING HOLISTIC DATA ANALYSIS

Very few programs have been written specifically to support holistic analysis. Fortunately, however, the needs of a holistic researcher are very similar to those of professionals in other fields where good programs have been written to help us do certain types of work. In addition to ordinary word-processing programs and database programs, three other types of more specialized programs may be useful: collaborative writing programs, software to support creative writing, and programs that support reflection. Two other types of software that have considerable potential, multimedia and hypermedia authoring software, have already been discussed.

Collaborative Writing Programs

Collaborative writing software helps researchers share documents and data with other participants in the study. Increasingly, more than one person is involved in the creation of a qualitative research paper. A growing number of papers have a large number of authors, including individuals who would have been considered "subjects" in traditional empirical research. This is an important aspect of some forms of participatory research (Heron, 1996). Programs in this category are sometimes called *groupware* because they allow more than one individual to work on a document without losing track of the difference between original material and revisions, and between the changes made by different authors. One example of collaborative writing software is CommonSpace, which costs about $75 per simultaneous user (Sixth Floor Media, 1999). It is available in versions for the Macintosh and Windows systems, and it runs on a local area network as well as networks that are connected to the Internet. (It can also be run on an individual computer with different authors taking turns.) For a detailed and positive review of CommonSpace, see Tucker (1998). For a comparison evaluation of CommonSpace and three other programs (Aspects, Daedalus, and Norton Textra) see Sands, 1996. (Note, however, that the distributors of CommonSpace supported this review.)

Another program that supports collaborative work is Aspects, which is distributed by Group Logic (1999). Aspects was designed primarily for use in classes where writing was a major focus (e.g., English composition, technical writing, foreign language, journalism, English as a second language). It runs on Macintosh computers and will work on a simple (and inexpensive) Appletalk network that is built into all Macintosh computers. (It will also work over the Internet so that participants do not have to be in the same room or on the same local area network.) Aspects supports collaborative writing and editing, and like CommonSpace, has a conferencing tool that lets participants discuss issues in both real-time (everyone is using Aspects at the same time), or asynchronously (collaborators participate at different times but the software keeps track of the discussion and where each person stopped). The conferencing features of CommonSpace appear to be somewhat more sophisticated than those available in Aspects.

A third program in this category is the Daedalus Integrated Writing Environment (The Daedalus Group, 1997). This program, like the other two already discussed, was created primarily for use in writing classes. Students can write, edit, and discuss their work in groups. Daedalus, like Aspects, was created specifically for use by a group of students who are taking the same course (although new versions of these programs have generally added Internet connectivity). Another program, The Collaborative Writing Engine, takes a different approach. It is Web-based, and users can be anywhere in the world as long as they have access to the World Wide Web. Participants can edit and revise a document and add comments about documents. The software is a project of Andrew Graham and Sigfried Gold (1999) at NCSA Emerging Technologies. The Web site for The Collaborative Writing Engine is at *http://www.shout.net/~sigfried/*.

Links to information on several other programs that support collaboration over the World Wide Web are available at *http://www.umr.edu/~rhall/research/Weblearning/Webconference.html* which is maintained by Richard Hall at the University of Missouri-Rolla.

Groupware can be used in two different ways by qualitative researchers. First, it can be used as a way of adding comments and interpretations to transcripts of interviews and observations. In all these programs comments can be kept separate from data, and the identity of the author of different comments and revisions can be maintained. Second, as noted above, groupware can support collaborative or participatory creation of both intermediate and final reports, and interpretations of the data. Both these general purposes are highlighted in a quote from Sixth Floor Media's home page. The quote also notes the difficulty of doing collaboration using traditional software such as word processors:

> There is a long, familiar tradition of annotation in most cultures, such as marginalia in ancient texts such as Torahs and illuminated manuscripts. These are early examples of annotation that are now echoed in a teacher's scribbled comments in the margins or between the lines of a student's paper. Each time an instructor or a peer writes comments on a paper, the paper is revised and the history of the document grows richer and more complex.
>
> Any computer user knows that conveniences of technology are unfortunately accompanied by frustrating constraints. In the life of the writer, word processing programs, conventional computer monitor sizes, and the ephemeral nature of on-screen text com-

> plicate the processes of making annotations and tracking of the history of the document. It's often easy enough to incorporate the comments of a single reader, but keeping straight the comments of some larger number of commentators can quickly become ungainly and difficult. In addition, incompatible hardware systems, the challenges of collaborating over networks, and the overwhelming and often baffling options that technology offers usually exacerbate a revision process that is already complicated enough. (Sixth Street Media, 1999, Online)

Software for Creative Writing

Software for creative writing makes writing plays, novels, and movie scripts easier. For some researchers, the idea of using this type of software to analyze qualitative data seems far-fetched at best. But if you take a holistic perspective and your purpose is to tell stories, then the idea seems much more natural and logical. Movies, plays, television programs, short stories, and novels all tell a story. Many programs are available for creative writing and a number of companies specialize in stocking this type of software and supporting those who use the programs. Scriptdude *(http://www.scriptdude.com)* is one such company. All the software mentioned in this section can be ordered on the Scriptdude Web site. Other distributors of this type of software include Ballistic Computing *(http://www.ballisticware.com/hollywood.html)*, Scriptware *(http://www.scriptware.com/content.htm)*, and the Writers Computer Store *(http://writerscomputer.com/cgi-bin/SoftCart.exe/store/brainstorming_reference_software.htm?E+writers)*

There are four basic types of creative writing software, and each type can serve certain functions in holistic analysis:

1. *Plot Brainstorming Programs.* Writing a story, whether for a television show, a movie, a play, or a qualitative research study, is a complex endeavor. The writer must think through and brainstorm many aspects of the story: story structure, plot, character development, dramatic relationships between characters, etc. There are a number of programs that support this critical aspect of writing a story. A popular one is Movie Magic Dramatica. It helps you build rich believable characters and a story line. Then it guides you through creating the scenes, beginning with creating an outline for a scene, followed by writing the scene. The program has a query system that lets you follow

the progression of themes and story elements through the story you are writing.

Like all the software discussed in this section, plot brainstorming programs were not created as qualitative research tools. However, holistic qualitative researchers may find these programs very useful because they support some of the same sorts of writing that emerge from holistic research–narratives and stories.

2. *Story Planners*. Some writers seem to write in a trench–with text strewn around them, they plow through to a destination that may or may not be known before they get there. Other writers step out of the trench regularly and look at the broader framework of the story they are telling. Then they get back in the trench and work on the text. This latter type of writer, someone who needs to develop a framework that will guide his or her writing, may find story planners useful. An example of a story planner is StoryVision. This program helps you create a visual map of your story. Diagrams lay out scenes and help you create them in standard formats in your word processor. StoryVision is particularly helpful when you are working on a complex story that has a number of layers or interrelated pieces. It has also been used by developers of interactive CD-ROMs to keep track of the different levels and problems in the video game or training material being produced. StoryVision has specialized functions such as the ability to create branching storylines–a choice at a particular point in the story determines which plot line is pursued. StoryVision helps writers create these complex multi-path materials and also helps them see the structure as it emerges through visual displays and maps that can be clicked to move to a particular part of the story.

Another program in this category is Writer's Block. If you are working with segments of material that will later be put into a single story or report, Writer's Block supports working on separate segments. They are put into "blocks" and the blocks can later be organized and merged into a story or document.

3. *Screenplay Programs*. Screenplay programs have basic word-processing features as well as specialized functions that support the creation of movie scripts, stage plays, television scripts, and commercials. Some, for example, let you organize dialog easily. Press the Tab key and the program expects the next word to be the name of a character who will speak. It automatically puts the character name in all caps and formats the dialog properly.

A popular program of this type, ScriptThing, was used to create the scripts for movies (including *Saving Private Ryan*) and television series (*The Practice* and *Ally McBeal*). Programs like ScriptThing and competitors such as FinalDraft, and ScriptWare are all stand-alone programs that have special features for specialized forms of storytelling. Other programs are add-ons. They are integrated with another program, usually a word processor, to give the generic program more specialized features. Script Wizard is a good example. It is designed to be used with Microsoft Word. Other programs in this category are somewhat specialized. *SidebySide*, for example, is designed specifically for the creation of scripts that have at least two columns. One column, for example, might contain stage directions and another might contain dialog. This two-column format, which can also be partially duplicated in Microsoft Word and other word processors, would be helpful to a researcher who wants to add comments and annotations in one column while the second column stores the text of the interview. *SidebySide* is an add-on program for use with Microsoft Word.

As qualitative researchers move beyond the traditional scholarly paper as the primary means of communicating research results, programs like ScriptThing and Script Wizard may appeal to a growing number of scholars. Many have provisions for adding annotations and notes, and a common feature is the ability to keep track of the plot and to move scenes and segments around with ease.

Software That Supports Reflection

Interpretive qualitative researchers tend to emphasize the thoughtful or reflective consideration of qualitative data. Like Collins, whose work was described earlier, they read and reread the data. Often they make notes about the data–questions, ideas, possible relationships or connections between one thing and another. They may also frequently ask others what they think or how they feel about tentative understanding. Several types of software can support this type of data analysis. The three types discussed already in this section can all be helpful. So can powerful word-processing programs that have features such as outlining, bookmarking, and multiple levels of headings. All these features let you structure and organize your data in ways that make it easier to move from one part of your data to another and try out ideas that emerge from your data analysis. Hypertext authoring programs are yet another type of software that supports reflective analysis.

SUMMARY

Several types of software are used by many qualitative researchers and accepted by most. These noncontroversial uses include computers as sources of information and aids to collaboration, electronic sources of data, and support for communicating results. There is considerable controversy over whether computers should be used in qualitative data analysis, and opinions range from decidedly negative to enthusiastically positive. Much of the difference can be explained by the paradigm the researcher adopts. Atomistic researchers tend to be very pleased with the computer software available for qualitative data analysis. Holistic researchers find less interest in the current collection of specialized programs, but there are many programs developed for other uses that can be adapted to holistic qualitative analysis.

REFERENCES

Chase, S. (1996). Personal vulnerability and interpretive authority in narrative research. In R. Josselson (Ed.), *Ethics and process in the narrative study of lives.* Thousand Oaks, CA: Sage.

Coffey, A., Holbrook, B., & Atkinson, P. (1996). Qualitative data analysis: Technologies and representations. *Sociological Research Online, 1*(1) [On-line]. Available: *http://www.socresonline.org.uk/socresonline/1/1/4.html*

Collins, P. (1998). Negotiating selves: Reflections on 'unstructured' interviewing. *Sociological Research Online, 3*(3) [On-line]. Available: *http://www.socresonline.org.uk/socresonline/3/3/2.html*

Graham, A., & Gold, S. (1999). The Web Site for The Collaborative Writing Engine. [On-line]. Available: *http://www.ncsa.uiuc.edu/edu/courses/spring98/assignments/collab_writing.html*

Group Logic. (1999). Group logic home page. [On-line]. Available: *http://www.grouplogic.com/GLI/about.html*

Herron, J. (1996). *Co-operative inquiry: Research into the human condition.* Thousand Oaks, CA: Sage.

Kelle, U. (1997). Theory building in qualitative research and computer programs for the management of textual data. *Sociological Research Online, 2*(2) [On-line]. Available: *http://www.socresonline.org.uk/socresonline/2/2/1.html*

Lee, R., & Felding, N. (Eds.) (1991). *Using computers in qualitative research.* London: Sage.

Lonkila, M. (1995). Grounded theory as an emerging paradigm for computer-assisted qualitative data analysis. In U. Kelle (Ed.), *Computer-aided qualitative data analysis: Theory, methods and practice*. London: Sage.

Miles, M., & Huberman, M. (1994a). *Qualitative data analysis*. Newbury Park, CA: Sage.

Miles, M., & Huberman, M. (1994b). Data management and analysis methods. In N.K. Denzin, & Y.S. Lincoln, (Eds.), *Handbook of qualitative research* (pp. 428-444). Thousand Oaks, CA: Sage.

Mulder, J. (1994). The mechanics of qualitative analysis. *Issues in Educational Research*, *4*(2), 103-108. [On-line]. Available: *http://cleo.murdoch,edu.au/gen/iier/iier4/942p103.htm*

Sands, P. (1996). *The Epiphany Project: Writing software comparisons* [On-line]. Available: *http://mason.gmu.edu/~epiphany/docs/comparison.html*

Sixth Floor Media. (1999). Sixth floor media home page. [On-line]. Available: *http://www.sixthfloor.com/index.html*

Smith, J. (1993). *After the demise of empiricism: The problem of judging social and educational inquiry*. New York: Ablex.

The Daedalus Group. (1997). The Daedalus group home page. [On-line]. Available: *http://www.daedalus.com*

Tucker, N. (1998). CommonSpace: A review. [On-line]. Available: *http://129.118.38.138/kairos/1.3/reviews/commonspace/openpage.html*

Leping Liu
Stuart A. Harrington
D. LaMont Johnson

What Influences the Quality of Research: Findings from an Editorial Management Database

In the field of education, educators and researchers conduct studies and investigations to discover facts or establish principles. These studies and investigations constitute educational research. The purpose of educational research is to improve the quality of education by increasing educators' knowledge of the art and science of teaching, and the process of learning (Suter, 1998). The value of educational research depends on not only what the researchers have discovered, but also to what extent they share their findings and knowledge with other educators. There are many ways to share findings, but the primary method is through writing reports, articles, or books (Kirt & Berger, 1993).

Effectively sharing knowledge is a two-way communication process

LEPING LIU is Assistant Professor, Department of Reading, Special Education and Instructional Technology, College of Education, Towson University, Towson, MD 21252. E-mail: lliu@towson.edu
STUART A. HARRINGTON is a Doctoral Candidate, Department of Curriculum and Instruction, University of Nevada, Reno, and Director of the Reno Learning Clinic, 855 West 7th Street, Reno, NV 89503. E-mail: stuarth@scs.unr.edu
D. LAMONT JOHNSON is Professor, Department of Counseling and Educational Psychology, College of Education, University of Nevada-Reno, Reno, NV 89557. E-mail: ljohnson@unr.edu

[Haworth co-indexing entry note]: "What Influences the Quality of Research: Findings from an Editorial Management Database." Liu, Leping, Stuart A. Harrington, and D. LaMont Johnson. Co-published simultaneously in *Computers in the Schools* (The Haworth Press, Inc.) Vol. 15, No. 3/4, 1999, pp. 53-65; and: *Information Technology in Educational Research and Statistics* (ed: Leping Liu, D. LaMont Johnson, and Cleborne D. Maddux) The Haworth Press, Inc., 1999, pp. 53-65. Single or multiple copies of this article are available for a fee from The Haworth Document Delivery Service [1-800-342-9678, 9:00 a.m. - 5:00 p.m. (EST). E-mail address: getinfo@haworthpressinc.com].

between authors and readers (Anderson, 1980; Becker, 1991; Beebe, 1993; & Blyler, 1995). From this perspective, authors should be able to disseminate knowledge effectively, and readers should be able to understand and use what they learn. However, there has frequently been a communications gap between researchers and readers caused by poor research design or writing (Kirt & Berger, 1993; Smeltzer & Thomas, 1994; Sanger, 1995; Rundall, 1996; & Charney, 1998). For example, Kirt and Berger (1993) summarized some common problems of research in the field of social work where some researchers tended to: (a) work on trivial problems, or apply trivial approaches to significant issues; (b) use inappropriate methods, (c) misinterpret results, or (d) bury their findings in technicalities. These problems prevent researchers from communicating effectively with readers and practitioners. Furthermore, the problems also impede communication among researchers.

It is likely that a similar communication gap exists between the producers and consumers of educational research. However, the factors that influence the dissemination of educational technology research have not been well documented. The primary purpose of this study is to identify research design and writing factors that limit the publication of articles within an educational technology journal, and to provide guidance to authors who wish to communicate through educational journals.

The identification of limiting factors was facilitated by the existence of an editorial management database (EMDB) for a refereed international journal in the field of information technology in education. The investigation summarized and analyzed the data in the EMDB. This database was created to manage the information from the authors, the journal reviewers, and the editorial staff. Although the database was created to enhance editorial record management, it also permitted research about the editorial process. For example, to investigate the common problems existing in current research, the investigators could quickly retrieve qualitative and quantitative editorial assessments on articles submitted for publication.

In the past, educators and researchers used index cards to manage research data (Soper, 1976; Apostle, 1987). The management of index cards was a job that required substantial expenditures of time to manually organize, sort, or retrieve information. With the development of computer technology, database applications replaced index cards. A review of literature showed that databases have been used in medical research (Thorogood & Armstrong, 1998; Johnson, 1996; Reynolds, 1994), clinical research (Walley & Mantgani, 1997; Jick, 1997), marketing research (Ezop, 1994), and educational research (Tucker, 1995; Clouser, 1998). However, the literature only described the use of databases as information resources for research, such as the UK General Practice Research Database for medical and clinic research, the Mark-Intel Online Database for marketing research, and ERIC, EBSCO, or Lexis-Nexis for educational research. There were few descriptions that focused on the practical issues of managing editorial data with a computerized database. Therefore, this study will also illustrate the process of using a database to manage editorial data, with the EMDB as an example.

PROCEDURES

Both qualitative and quantitative data were analyzed in this study. First, the qualitative data were summarized and categorized to identify major problems that influenced the quality of research. Second, the quantitative data were analyzed to determine which of the major problems could be the important variable(s) that predicted the probability that an article would be published. A logistic regression analysis was performed to estimate this probability.

The investigation was based on information stored in the EMDB of an international journal that receives submissions from authors throughout the world, including the United States, Canada, South America, Europe, Australia, and Asia. The authors consisted of school teachers, graduate students, professors, and experts in the field of using computer technology in education. Therefore, it was assumed that the information was representative of articles submitted to journals in this field.

QUALITATIVE ANALYSIS AND FINDINGS

Journal reviewers' comments were sorted into six categories: (a) theme and focus, (b) research design, (c) measurements or instruments, (d) research findings, (e) references, and (f) writing. These categories follow the model suggested by traditional research textbooks (Gay, 1992). However, since comments from reviewers were occasionally difficult to categorize, the most commonly identified problems within each category were summarized. Each field was searched to establish the frequency of these common problems. If over 30% of the reviewers identified an issue as a problem, that issue was defined as a "frequently addressed" problem.

Theme and Focus

The qualitative analysis sought to identify frequently addressed problems with regard to theme and focus. An appropriate theme should be based on the literature review and focus on an area that can produce meaningful outcomes. The following are some major problems in determining the theme and focus of the study.

Some themes have been studied repeatedly over many years, with similar findings, leading to similar suggestions. These themes may have been interesting in 1979, but not in 1999. If the article could not provide new theoretical or practical insights, it would not be publishable. If the research focused on cutting-edge technologies, produced meaningful new findings, or suggested new approaches of practical application in education, the qualitative analysis suggested that such articles would be revised and published unless the article had serious research design problems.

Second, some authors' focus was narrow and shallow, and contributed no new insights to the literature. For example, an article that assessed gender differences in computer technology use, by simply counting the number or frequency of males and females who used the computer, were considered to have a trivial or shallow focus. Such papers were not considered well researched, even though they were sometimes well written.

A thorough literature review will identify frequently studied topics, research trends, and unsolved problems. The literature review may also prevent researchers from wasting time on themes that are not researchable.

Third, article titles were inconsistent with the content. While this problem may seem unimportant, the inconsistency between an article's title and content leads to misunderstandings between authors and reviewers. Four inconsistencies in this regard were noted. First, some articles were totally disjointed, and the content had little to do with the title. Second, some articles had ambiguous titles that did not identify what was actually discussed, or the title misled the reviewer about the article's content. Some articles had very long titles that extended for three or more lines, or consisted of more than two separate sentences or questions. Third, other article titles contained significant issues, but the text discussed trivial problems. As a result of this inconsistency, journal reviewers felt that the arguments were weak, or the results were insufficient. Fourth, the article titles contained inappropriate wording or terms (e.g., using quantitative terms for a qualitative study). Writing an article without a concise and precise title consistent with the content is like "drawing a dragon without drawing its eyes" (an old Asian aphorism). However, revising a title is easily accomplished.

Research Design

After evaluating the theme and focus, reviewers turned their attention to the research design. Several research design problems were identified in the EMDB.

First, there were several common sampling problems. Sampling is "the process of selecting a number of individuals for a study in such a way that the individuals represent the larger group from which they were selected" (Gay, 1992, p. 123). Some studies mentioned a "randomly selected sample" without describing the selection process. Some samples had a seriously skewed male-female ratio (e.g., 146 males to 16 females). Some samples were too young to be representative of the students who were learning a particular technology. In some studies, sample sizes were inconsistently reported. In all these cases, the studies failed to provide either an adequate description or explanation of the sample.

The purpose of sampling is to gain information about a population. Some studies did not even try to identity and define the population. Research was valuable only if it was generalizable, but results and findings obtained from such samples were not generalizable. Therefore, the value of the study was greatly diminished.

The second problem was weak design. Some studies were conducted as qualitative and written as quantitative research, or vice versa. This inappropriate design hides the potential value of a study and disguises important findings. Another research design problem noted was the use of inappropriate methods for data analysis. A statistical analysis method was appropriate only when it was consistent with the research design. For example, it is inappropriate to use a *t*-test for a pretest-posttest control group design; analysis of covariance is the preferred method for such comparisons (Campbell & Stanley, 1963).

The third problem was that some articles were cast in research terms and format, but they were not research in nature. For example, describing a one-week course or a Web-page design does not constitute research. Whatever quantitative findings these papers offered were difficult to interpret and generalize.

Measurements and Instruments

Instrumentation was an important issue that influenced the validity and reliability of research findings. According to the information stored in the EMDB, the major instrumentation problem was that studies failed to provide evidence to substantiate the validity of their instruments. Some studies did not report the validity and reliability of their instruments. If there was no evidence to demonstrate that an instrument consistently measured what it was supposed to measure, the measurement results were questionable. Some studies used two different instruments to measure the same variable for two groups without reporting the consistency of the instruments. If the tests administered to two groups measured different variables or different levels of knowledge, the author was not justified in discussing "differences" between the two groups. Some measurements were based on surveys with very low return-rates (e.g., lower than 40%). If a survey return rate was lower than 40%, the generalization to the population from which the sample is drawn is unwarranted (Babbie, 1994). Finally, some instruments were inappropriate for the sample (e.g., the test was too difficult for subjects of that age group). Conclusions based on such results were also questionable.

Findings and Conclusions

The findings and conclusions of authors were also important components for reviewer critique. According to data in the EMDB, there were two major problems in developing conclusions.

First, studies were often inconclusive. Some had important findings but failed to draw a conclusion that explained how these findings could make contributions, theoretically or practically, to the literature. Often, articles described or repeated the research findings in the conclusion section. This, of course, weakened the study by not telling the reader how the findings might be used. In drawing conclusions, reviewers preferred strong arguments that used research findings to support conclusive statements.

Second, conclusions were often weak. Inspection of the EMDB suggested that there were two major causes for weak conclusions. The findings were either minimal or too ambiguous to support a strong argument. The inconclusive, weak, or ambiguous studies were most often associated with poor research design (as discussed previously).

Conclusions that were unrelated to the introduction were also considered weak. The conclusion provides closure to the problems, purposes, questions, and hypotheses of the research posed in the introduction. The conclusion states what the study contributed to the field and how the findings could improve current practice or guide further research.

References

There were three major problems concerned with appropriate use of references. First, often references were not up-to-date. In the field of educational technology, where change is a daily occurrence, current references are essential. Second, references were either insufficient or superfluous. Some studies did not provide sufficient background information or knowledge in the area, and some literature was unrelated to the topic of the article. The third problem was with the interpretation of the literature. Some literature reviews failed to create an integrated picture of current theory and knowledge, but rather consisted of a series of single-sentence summaries of articles. Unlike problems with research design, problems with the literature review can be remedied with revision.

Writing

Writing deficiencies were most frequently addressed in the EMDB. Problems with wording, mechanics, organization, and transition were common and are beyond the scope of this article. However, journal reviewers offered some suggestions that might be helpful to prospective authors.

1. Write for the reader. For example, authors need to help the reader gain a simple, conceptual understanding of the techniques, concepts, or complex procedures of the study (e.g., statistical procedures or a subtle interpretation of the results).
2. Use terms consistently. For example, after naming the dependent and independent variables in the methods section, use those names consistently throughout the article.
3. Use verb tense consistently. Reviewer's comments suggested very clear guidelines: use past or present perfect tense in the literature review and method sections; use past tense for discussing results; and use present tense for discussing findings and conclusions or for explaining figures and tables.
4. Vary paragraph length to keep reader interest. For example, avoid paragraphs that exceed one page in length.
5. Avoid writing the article as if "you were forced to write a course assignment."

Reviewers also noted that some submissions looked like first drafts rather than final copy. The writing process requires much refining prior to submission.

In this section, some typical problems related to the quality of research were discussed from a qualitative perspective. The major problems are summarized in Table 1. These problems will next be examined from a quantitative perspective.

QUANTITATIVE ANALYSIS AND FINDINGS

Methods

The findings of the previous section demonstrate that the quality of research articles, with regard to the particular journal under study, is related to the quality of: (a) theme and focus, (b) research design, (c) measurements or instruments, (d) research findings, (e) references, and (f) writing. To what extent these six variables are related to the probability of publication will now be considered.

To estimate the probability that an article will be published, a logistic regression was performed. Logistic regression is a form of statistical modeling that is often appropriate for categorical outcome vari-

TABLE 1
Research Quality Analysis Summary

Quality Items	Problems or Weaknesses
Theme and Focus	Not up-to-date
	Narrow and shallow
	Inconsistent with content
Research Design	Sampling
	Methods
	Generative
Instrument	Invalid
	Inconsistent
	Low return rate
Findings and Conclusions	Inconclusive
	Weak conclusions
	Irrelevant with introduction
References	Not up-to-date
	Insufficient or superfluous
	Interpretation
Writing	Term consistency
	Verb consistency
	Paragraph length varies

ables. It describes the relationship between a categorical response variable and a set of explanatory variables (Stokes, Davis, & Koch, 1996; Neter, Kutner, Nachtsheim, & Wasserman, 1996). The response variable is usually dichotomous; typically, the two outcomes are *Yes* and *No*. The explanatory variables in logistic regression can be categorical or continuous. The purpose of a logistic regression analysis is to detect the relationship between the explanatory variables and the probability of the response variable to have the outcome *Yes* (or *No*). For example, for a response variable "teenager smoking," the outcome 1 can indicate "smoking," and 0 "non-smoking." An explanatory variable "gender" can have the outcome 1 as "male," and 0 as "female." The logistic analysis would detect whether gender relates to the probability for a teenager to be a smoker.

Variables

In this study, the explanatory variables were the six research quality items. They were all categorical variables with the outcomes 1 and 0. One (1) indicated that the variable met the quality criteria, and 0 indicated that the variable did not meet the quality criteria. The response variable was article status (publication decision), with the outcome 1 as *published*, and 0 as *rejected* (as shown in Table 2).

It was hypothesized that at least one of the six variables would have a relationship with, and could be used, to predict the probability of an article being published.

The Logistic Analysis and Results

One hundred and two articles submitted during the last three years were randomly selected from the EMDB. Each article was selected by matching its log number with the last two digits of a calculator-generated random number. Forty articles out of the 102 were actually published.

TABLE 2
Variables in the Logistic Regression

Variables E: explanatory variable R: response variable	1	0
Article Status (R)	Published	Rejected
Theme/Focus (E)	New and Interesting	Not Up-to-Date and Non-Researchable
Research Design (E)	Appropriate and Valid	Poor and Invalid
Instruments (E)	Valid and Reliable	Invalid and Unreliable
Findings/Conclusions (E)	New and Generative	Weak and Irrelevant
References (E)	Sufficient and Updated	Insufficient and Not Updated
Writing (E)	Good	Not Good

The analysis consisted of identifying a model that best fit the data and analyzing the data. The first logistic analysis produced a model that consisted of all six explanatory variables. In this initial model, only the Research Design variable was significantly ($p < 0.0117$) related to the response variable, Publication.

In the second logistic analysis, Research Findings, Instrument, and Research Design variables were retained in the model. Research Findings and Instrument variables were included in the second model because they approached significance ($p < 0.1651$, and $p < 0.1104$, respectively) in the first model. It was hoped that including them in the second model would contribute significantly to the prediction of publication. However, this was not the case. In the second model, Research Design was the only variable that contributed significantly ($p < 0.0006$) to the prediction of the response variable. A third logistic analysis was performed which included only the Research Design variable, and it was significant ($p < 0.0202$).

To determine the best model to use, we compared the AIC (Akaike Information Criterion) values which is a goodness-of-fit measure. Since small AIC values indicate a more desirable model, the second model that contained the Research Findings, Instrument, and Research Design variables was selected (AIC value of 115.214 as compared to 117.082 and 135.061). This model indicated that the combination of these three variables would provide an explanation of the relationship between the research quality variables and the probability of an article being published.

As shown in Table 3, the Research Design variable was a significant ($p < 0.0006$) predictor of article publication. Another useful value in this table that helped explain the probability of the response variable

TABLE 3
Logistic Analysis Outputs

Variable	DF	Parameter Estimate	Wald Chi-Square	p > Chi-Square	Odds Ratio
Intercept	1	−2.7753	14.3821	0.0001	
Findings	1	0.2575	0.4722	0.4920	1.294
Instruments	1	1.4184	1.5658	0.2108	4.130
Design	1	2.7162	11.9086	0.0006	15.123

was the Odds Ratio. If the Odds Ratio was larger than 1, the probability of an article being published would increase and if smaller than 1, the probability would decrease. In this model, the Odds Ratio for the Research Design variable was 15.123. This meant that an article that met the Research Design quality criterion would be 15.123 times more likely to be published than an article that failed to meet the criterion.

The analysis summarized in Table 3 indicated that the Research Design variable was the most important predictor of an article being published. Even though the Research Findings and Instrument variables did not make a significant contribution to the model, the Odds Ratios for these variables suggested that the probability of publication increased when the article met those quality criteria as well.

SUMMARY

In this paper, we investigated the major problems with research and research articles for an academic journal. Our findings suggest that research quality problems fall into six categories: (a) theme and focus, (b) research design, (c) measurements or instruments, (d) research findings, (e) references, and (f) writing. Among these categories, Research Design was the most influential in determining whether or not an article was published.

Conducting research is always a learning process and represents a good opportunity for evaluation and improvement. The study offers guidance to prospective researchers since those who devote more time to research design, instrumentation, and writing conclusions will increase their chance of communicating with their audience.

REFERENCES

Anderson, P. (1980). The need for better research in technical communication. *Journal of Technical Writing and Communication*, *10*, 271-282.

Apostle, L. (1987). *Personal file management for the health sciences* Washington, DC: George Washington University. (ERIC Document Reproduction Service No. ED 281 560)

Babbie, E.R. (1994). *The practice of social research* (7th ed.). Belmont, CA: Wadsworth.

Becker, B.J. (1991). The quality and credibility of research reviews: What the editors say. *Personality and Social Psychology Bulletin*, *17*, 267-272.

Beebe, L. (Ed.). (1993). Professional writing for the human services. Washington, DC: NASW Press (National Association of Social Workers).

Blyler, N. (1995). Research as ideology in professional communication. *Technical Communication Quarterly, 4*, 285-313.

Campbell, D.T., & Stanley, J.C. (1963). *Experimental and quasi-experimental designs for research* Boston, MA: Houghton Mifflin.

Charney, D. (1998). From logocentrism to ethnocentrism: Historicizing of writing research. *Technical Communication Quarterly, 7*, 9-33.

Clouser, J. (1998). A magnifying class. *Civil War Times Illustrated, 37*(4), 10-11.

Ezop, P. (1994). Database marketing research. *Marketing Research, 6*(4), 34-42.

Gay, L.R. (1992). *Educational research: Competencies for analysis and application.* New York: Macmillan.

Jick, H. (1997). A database worth saving. *Lancet, 350*(9084), p. 1045.

Johnson, R.C. (1996). Research database has a mind to map the brain. *Electronic Engineering Times, 895*, 33-35.

Kirt, S.A., & Berger, R.M. (1993). Improving research writing. *Social Work Research and Abstracts, 29*(4), 3-5.

Neter, J., Kurner, M.H., Nachtsheim, C.J., & Wasserman, W. (1996). *Applied linear regression models.* Chicago, IL: Donnelley & Sons.

Reynolds, T. (1994). Linked databases help researchers track costs, care patterns, outcomes. *Journal of National Cancer Institute, 86*(3), 168-172.

Rundall, T. (1996). Conducting and writing research reviews. *Medical Care Research and Review, 53*(1), 132-146.

Sanger, J. (1995). Five easy pieces: The deconstruction of illumination data in research writing. *British Educational Research Journal, 21*(1), 89-98.

Smeltzer, L.R., & Thomas, G.F. (1994). Managers as writers: A meta-analysis of research in context. *Journal of Business and Technical Communication, 8*(2), 186-212.

Soper, M.E. (1976). Characteristics and use of personal collections. *Library Quarterly, 46*(4), 397-415.

Stokes, M.E., Davis, C.S., & Koch, G.G. (1996). *Categorical data analysis using the SAS system.* Cary, NC: SAS Institute.

Suter, W.N. (1998). *Primer of educational research.* Boston: Allyn and Bacon.

Thorogood, M., & Armstrong, B. (1998). Mortality in people taking selegiline: Observational study. *British Medical Journal, 317*(7153), 252-258.

Tucker, R.E. (1995). The ideological effects of corporate database centered research: A response to Harris. *Argumentation and Advocacy, 32*(1), 46-50.

Walley, T., & Mantgani, A. (1997). The UK general practice research database. *Lancet, 350*(9084), 1097-1100.

Stuart A. Harrington

Maintaining Research Documents with Database Management Software

CARD FILES

One of the routine tasks of research is making and managing notes, reprints and other documents. After a short time, these documents begin to accumulate, gradually taking over our desk, the in and out basket, shoe boxes, folders, corners, and then the floor of our office. As the piles rise, it becomes increasingly difficult to locate a specific note in a reasonable amount of time. Enter the card file.

The card file allows the notes to be organized, stored, and retrieved in some systematic fashion (Apostle, 1987). The system that is imposed is the key to storage and retrieval. For example, if we have a set of notes on publications, these notes may be arranged alphabetically, by the author's last name. While it takes extra time to store our notes alphabetically, it greatly speeds the process of finding a specific note. Unfortunately, the alphabetical arrangement is of little utility if we forget the author's name or if we want to find all the notes on a particular topic. It can be frustrating to know that the reference or the note you need is within arm's length, but unavailable (Singer, 1979). Enter the database management program.

STUART A. HARRINGTON is a Doctoral Candidate, Department of Curriculum and Instruction, University of Nevada, Reno, and Director of the Reno Learning Clinic, 855 West 7th Street, Reno, NV 89503. E-mail: stuarth@scs.unr.edu

[Haworth co-indexing entry note]: "Maintaining Research Documents with Database Management Software." Harrington, Stuart A. Co-published simultaneously in *Computers in the Schools* (The Haworth Press, Inc.) Vol. 15, No. 3/4, 1999, pp. 67-80; and: *Information Technology in Educational Research and Statistics* (ed: Leping Liu, D. LaMont Johnson, and Cleborne D. Maddux) The Haworth Press, Inc., 1999, pp. 67-80. Single or multiple copies of this article are available for a fee from The Haworth Document Delivery Service [1-800-342-9678, 9:00 a.m. - 5:00 p.m. (EST). E-mail address: getinfo@haworthpressinc.com].

Database management programs can function like flexible card files, easing the process of storing and retrieving the notes about any topic. Managing notes is a task that is ideally suited to computerization. All that is required is a little knowledge of database management software and a plan. This paper will review the literature on personal filing systems, introduce the basic process of database management, and offer a plan for managing your research notes.

PERSONAL FILING SYSTEMS

Personal files are one of the researcher's basic tools. Researchers use these files when they perform research, write about their research, and communicate with their colleagues (Wanat, 1985). Efficient organization of these files can greatly increase the researcher's productivity, if the collection meets two criteria (Jahoda, 1970). First, the collection must be complete enough so that it is worth organizing and indexing. Second, the collection must be one that is used frequently.

The evidence suggests that the personal files of many researchers meet these two criteria. Researcher's files often cover specific topics thoroughly, and the files are often extensive (Soper, 1976). Jahoda, Hutchins and Galford (1966) found that two thirds of the researchers they surveyed kept indexed personal files. They also found that about half of the researchers that they surveyed used their indexes daily and another 39% used their indexes weekly. Furthermore, Soper (1976) found that researchers preferred to cite sources from their personal collections, possibly because these collections are smaller, less complex, and more accessible than institutional libraries. These findings suggest that many researchers have personal files that are useful enough and large enough to warrant designing a personal filing system.

The process of designing a personal filing system has been thoroughly discussed in "how to" articles by researchers and clinicians. These articles explain the methods that these individuals use to manage their personal files. Some articles focus on manual index card systems (Apostle, 1987; Holmes & Gentry, 1979; Ringel, 1980; Singer, 1979), while others focus on computerized indexing systems (Beckley & Bleich, 1977; Bertrand & Bader, 1980; Calvin, 1972; Lipsett & Haycock, 1982; Rowel & Utterback, 1984). These discussions tend to present idiosyncratic approaches to maintaining a personal file that vary widely because they reflect the unique characteristics of each individual's situation.

While the characteristics of personal files vary depending on the needs of the researcher, personal files all share several characteristics. They contain documents that are familiar to the researcher. The documents have been carefully evaluated and selected for relevance. Finally, they are close at hand, forming a private collection that reflects the unique interests of that researcher (Wanat, 1985).

When designing a personal filing system, the researcher should carefully consider what kinds of materials will be managed. Books, journal issues, and copies of journal articles may comprise the majority of most filed materials (Soper, 1976), but other types of materials may also be important. The materials in a personal file might include (a) reprints or preprints, (b) technical reports, (c) papers delivered at professional meetings and related handouts, (d) correspondence, (e) lecture notes, (f) original articles, (g) references to books, (h) proposals, (i) manuals, (j) specifications, and (k) financial reports (Apostle, 1987; Wanat, 1985; Wright, 1992).

These research materials may be arranged in a variety of ways: by author, by alphabetical subject, by classified subject, or by accession number (Apostle, 1987). However the materials are arranged, it is best to choose a method of arrangement that permits rapid retrieval. For example, filing the materials by subject while you are researching a specific topic may speed retrieval. When that task is completed, these materials may be refiled for long-term storage in a personal archive under the author's name or accession number. Regardless of the method that you choose, it is important to store documents in an organized manner (Apostle, 1987).

As Wanat (1985) explains, the organizational structure of a personal filing system should be designed to meet several specific goals. First, the system should prevent information from being lost. Second, the system should permit flexible searching, not just by author or subject, but by any criteria useful to the user. Third, the system should facilitate the organization and evaluation of materials to enhance research or professional practice. Finally, the system should make the writing process easier and more productive.

To accomplish these goals, Wanat (1985) recommends that the designer of a personal filing system anticipate the following problems:

1. The needs of the researcher and the focus of the discipline will change. Therefore, employ a system that can be modified or adapted easily.
2. Even with rigorous weeding of the collection, the amount of material will grow. Therefore, plan for growth by employing a system that will accommodate large numbers of items.
3. As the collection grows linearly, the difficulty of finding any specific item will increase exponentially. Therefore, employ a system that permits flexible searches.
4. The amount of effort expended during filing is inversely related to the ease of finding a given item later. Therefore, expect to expend effort during the initial filing process.
5. If the system is difficult to use, it will not be used. Therefore, develop the simplest system that will satisfy your needs.

Database management software may provide the optimum solution to these problems. Database management software is adaptable; spacious; permits fast, flexible searches; and is relatively simple to use.

DATABASE MANAGEMENT PRIMER

There are numerous database programs on the market, and explaining the operation of these programs is beyond the scope of this article. Users must seek specific information about their program from that program's documentation. This database primer will only define a few terms that are essential for this discussion.

A database is an organized collection of information. This collection exists as one or more database files. When a new database is created, the user must name the new file.

After the new database file is named, the user will be asked to define fields that will contain specific types of information. Fields in a database are individual categories of information, such as author, date, and title. A field definition specifies the type of information that can be stored in that field. For example, text fields and memo fields can contain any characters you can type, but numeric fields will only accept numbers, date fields will only accept dates, and time fields will only accept times.

To define a field, select the field data type and select the maximum number of characters that may be entered into that field. Once these fields are defined, the database file is created, and the program will permit you to begin working with the new database.

The next step is to enter information. There will be a menu option for creating a new record. Select that option. For a journal article, enter the author's name, the year of publication, and title in the first three fields. At this point, you have created your first record.

Each record can be thought of as a single card in a card file. Each time you enter information on a new document, you create a new record. A database file is a set of records, each record consists of a set of fields, and each field stores information (Approach Software Corporation, 1994). The fields in a record hold the notes on one document.

FIELD GROUPS

The definition of fields may have long-term consequences, yet it is one of the first decisions that the user must make. Field definitions set limits on the use of the data, and the limits will prove both beneficial and detrimental at some point in the future. However, database management programs are flexible enough so that most difficulties can be overcome. With that caution in mind, I will make some specific suggestions about field definitions based on my experience.

Reference Fields

The first group of fields that I will discuss is based on the requirements for American Psychological Association (APA) style references (American Psychological Association, 1994). The reference fields include: *Author*, *Date*, *Title*, *Journal/Vol/Pages*, and *Publisher*. The reference information is entered into separate fields for three reasons. First, this arrangement permits simple, intuitive searching and sorting. Second, many database programs (Dbase, Paradox, FoxPro, and Access) limit text fields to 254 or 255 characters (Approach Software Corporation, 1994), and some complete references exceed that length. Third, memo fields, which have no fixed length, are unsuitable because they cannot be sorted. Therefore, separate text fields were used for each element in the reference (see Table 1).

The first element in the reference group is the *Author* field. This is

TABLE 1
Suggested Fields for Research Notes

Name	Type	Size
Reference Fields		
Author	Text	254
Date	Text	30
Title	Text	254
Journal/Vol/Pages	Text	254
Publisher	Text	50
Abstract	Memo	
Search Fields		
Action	Text	40
Purpose	Text	254
Personal Notes		
Conclusions	Memo	
Limitations	Memo	
Parameters		
Age/Grade	Text	40
N	Text	40
Method	Text	254
Instrument	Text	254
Independent Variable	Text	254
Dependent Variable	Text	254
Control Variable	Text	254
Validation Measures	Text	254

defined as a text field with 254 characters. While most author data is much less than 254 characters in length, there are a few articles with many authors.

The information in the *Author* field is entered in reference format. For example, the authors Ronald P. Carver and Susan W. Clark would be entered as: Carver, R. P., & Clark, S. W. For a book, this field holds the name of the book's author, and for a chapter in an edited book, this field holds the name of the author of the chapter.

The second element in the reference group is the *Date* field. A text field is used to hold date information because not all the dates used in references are actually dates. The date of an article may be 'in press' when you create your record. If the *Date* field were defined as a date type field, the program would not allow the words 'in press' to be entered.

The third element in the reference group is the *Title* field. This field is defined as a 254-character text field to accommodate long titles. The title should be entered exactly as it will appear in a reference, with correct use of capitalization, and followed by a period. For example, the article entitled "Levels of Processing: A Framework for Memory Research" would be entered into the field as: Levels of processing: A framework for memory research. The title of a book or the title of a chapter in an edited book could also be entered into this field.

The fourth element is the *Journal/Vol/Pages* field. As the name suggests, this field stores several pieces of information. This field is defined as a 254-character text field also. This length would not be required if the only information entered here was journal name, volume number, and the pages of the article. However, this field may also store the name of the editor, book title, and pages of a chapter in an edited book such as the location of a chapter by Bjork: In H. L. Roediger III & F. I. M. Craik (Eds.), Varieties of memory and consciousness (pp. 309-330). The description of the editor, book title, and page entry can be fairly long, hence the need for a large text field. This field can also be used for other information that follows the title in the reference, such as the source for government reports or document deposit service information, such as Educational Resources Information Center (ERIC) references (Chesley, 1989).

The final element of the reference group is the *Publisher* field. An example of a publisher entry is Hillsdale, NJ: Erlbaum. This field is used for publishers of books and is left blank for journal articles.

Abstract

A field that is closely related to the reference group is the *Abstract* field. This field is defined as a memo type field. Abstracts frequently exceed the maximum character limits of text fields so a memo field is the best field type option.

Search Fields

The reference and abstract fields contain information that is taken directly from the articles as they are reviewed. These next two fields are created by the database user to mark or categorize the record for future searches.

The first search field is called the *Action* field. This field is defined as a text field with a length of 40 characters. This field identifies the next action that needs to be performed on this article. For example, the entry "library" tells me that this is a reference that I need to locate in the library. After the reference is located and reviewed in the library, "library" might be changed to the call number of the reference, so I could find the reference quickly in the stacks the next time it is needed.

If I choose to copy the article for future reading, I change "library" to "update" to remind myself to read the article and update the record with my observations. Once a copied article is read, I change the *Action* field to "filed" and follow the word "filed" with the location where the article is filed. For example, if the article is filed in a binder titled phonology, I will enter "filed phonology."

If the article is unavailable, I may decide to order the article from a document delivery service. If I plan to place an order, I change the *Action* field to "order." If I wish to retain the record but take no further action, I enter "NA" for not available. If I decide the article is not important enough for future reference, I delete the entire record.

The second search field is called the *Purpose* field. The *Purpose* field holds key words that describe the subjects covered in the document. For example, for an article by J. K. Torgesen entitled "The study of short-term memory in learning disabled children: Goals, methods, and conclusions," I have entered two terms in the *Purpose* field: "LD STM." My purpose for making and keeping a copy of this article is that it summarizes information on the short-term memory (STM) of children with learning disabilities (LD). As these terms suggest, I make extensive use of acronyms to minimize typing.

It is important to make a list of the key words that you use in the *Purpose* field. It may also be wise to consider using a specialized thesaurus or diagnostic manual instead of creating your own terms. A thesaurus provides a comprehensive set of cross-references, synonyms, singulars and plurals, and suggests appropriate word order. For example, educational terms can be located in the *Thesaurus of ERIC Descriptors* (Houston, 1995) and psychological terms can be found in the *Thesaurus of Psychological Index Terms* (American Psychological Association, 1997). If diagnostic categories are essential in your field, the key words might be obtained from a diagnostic manual, a medical index (Ringel, 1980), or even a basic textbook (Singer, 1979).

Personal Notes

The next two fields contain personal notes on the article. The first of these note fields is called *Conclusions*. Observations and conclusions about the article are typed into this field.

The second note field is called *Limitations*. Significant limitations or design flaws in the study are typed into this field. Strengths and potentially useful ideas are typed into the *Conclusions* field and weaknesses and limitations are typed into the *Limitations* field. There is no strong reason to separate notes into two fields. Separating the notes reflects my approach to making and using these notes, and this approach may not be appropriate for other users.

Parameters

The parameters group of fields is designed to store information that describes the details of a study. The parameter fields are: *Population, Age/Grade, N, Method, Instrument, Independent Variable, Dependent Variable, Control Variable*, and *Validation Measures*. The titles of these fields describe the information that is entered into each field.

The use of the singular in these field names is not meant to suggest that an article only reports on a single population, uses a single instrument, or employs a single independent variable. Some articles report on three or four separate studies, use several instruments, or employ multiple independent and dependent variables. All of these details can be entered into each of these fields.

Organizational Schemes

While it is important to have an organizational scheme, it does not need to be the scheme described in this article. The fields described herein reflect my uses and are offered as an example. The user may wish to add other fields and eliminate some of these fields. Most database programs will allow for adding and deleting fields at any time. However, be cautious when deleting fields that contain information. When fields are deleted, the data in those fields are lost.

BROWSING THE DATABASE

If every field in a record were completely visible, the fields would fill several screens. Database management programs permit the user to determine which fields will be visible and how much of each field will be shown. Removing a field from view does not delete the data in that field, it just hides it from view. Consequently, the user can develop views that arrange the data to suit different purposes.

The first view in this database is the data entry screen. This screen places the reference fields and search fields at the top, with the notes next, and the parameters fields at the bottom. This arrangement is appropriate for beginning new records. If a useful source is identified, I create a new record and copy the information from the reference section into the reference fields of the new record. Then, I type "library" in the *Action* field to indicate that I plan to obtain this article from the library. I usually enter at least one key word in the *Purpose* field that suggests the purpose that I expect this source to serve.

Another view of the data puts more emphasis on the parameters and personal notes groups of fields. While reference fields are again at the top of the screen, the parameters and personal notes groups of fields are moved closer to the top of the screen, and the *Abstract* field is moved to the bottom. This arrangement is appropriate for detailed note taking while reading an article.

DATA ENTRY

Up to this point, data entry has been described as a manual process. In most cases, the parameters and personal notes fields are filled in from the keyboard. These fields hold personal observations or bits of data that must be extracted from the text of an article. However, the reference fields contain data that are either stored in other databases or in printed form. The information in these fields may be acquired with other, less labor-intensive, methods.

Information that is stored in databases, either online or on CD-ROM, can be captured in a text file. For example, an abstract on ERIC can be downloaded to a text file. Printed text can be copied and scanned into a text file. The text file is then loaded into a text editor. Once in the editor, data can be copied to the clipboard, and then pasted into the appropriate field of the database record. While this process requires many steps, it is not as difficult as typing all of this information.

These manual steps can be eliminated by employing bibliography-formatting programs that are designed to import references from electronic sources (Hoke, 1994) such as CD-ROM databases, reference research services, and Internet-accessible online libraries (Brown, 1994). The bibliography-formatting programs can import bibliographic information, store the information, assemble an appropriately formatted bibliography, and export the bibliography to word-processing programs. Some of the more flexible bibliography-formatting programs are also capable of performing some general database management tasks, identifying citations in a manuscript, and automatically assembling the bibliography of references cited (Stigleman, 1994), or connecting directly to the Internet (Stigleman, 1996). (For a review of bibliography formatting software see Stigleman, 1994 and 1996.) By automating the acquisition and creation of bibliographic data, bibliography-formatting programs can minimize the data entry error and eliminate routine tasks.

Reports

One of the great strengths of database management programs is their ability to locate, format, and print sophisticated reports. During the literature review process, a series of references is entered into the database. These references are acquired through reading the literature, from computerized databases, and from indexes such as the Reader's Guide to Periodical Literature, Dissertation Abstracts International, and Psychological Abstracts (Gay, 1992). After these references are entered into the database, a Library Search Report can be printed out to help locate articles and books in the library stacks.

To create a Library Search Report, find all the records that have "library" in the *Action* field, then sort them by the *Journal/Vol/Pages* field. Figure 1 presents the first two listings from a typical Library Search Report. When printed, this report lists all of the essential information for locating a reference in the library except the call number. Furthermore, the information is alphabetically arranged by journal. If you have several references to locate in a journal, they are printed together on the Library Search Report. Approximately 10 references print per page, so the report makes a compact list to carry into the stacks while you locate books and articles.

The database manager can also produce tables of selected information that emulate 3-by-5 cards. For example, you could print a "card" with the *Author* and *Date* field at the top. The body of the "card" might consist of your notes, which are located in the *Conclusions* and

Limitations fields. Furthermore, you could search the *Purpose* field for only those records that refer to a topic, such as short-term memory (STM). The final report will consist of a series of "cards" that each contain notes on short-term memory and enough information to create a citation.

A similar report could be created with *Author*, *Date*, and the parameters group of fields (see Figure 2). If you want to identify all of the experimental studies on a topic, enter "experimental" as the search criteria in the *Method* field. Once these records are identified, you could sort them by *Date* to review the parameters of all experimental studies in your database from oldest to most recent.

The types of reports that you can generate are limited only by the

FIGURE 1
Library search report.

Author	**Date**	**Title**		
Spring, C.	1976	Encoding speed and memory span in		
Journal/Vol/Pages			**Action**	**Purpose**
Journal of Special Education			Library	STM
Author	**Date**	**Title**		
Chi, M. T.	1976	Short-term memory limitations in		
Journal/Vol/Pages			**Action**	**Purpose**
Memory and Cognition, 4, 559-572			Library	STM

FIGURE 2
Parameters report.

Author		**Date**	
Spring, C., & Capps, C.		1974	
Method	**Population**		**Age/Grade**
Experimental	Normal and poor readers		7-13
yrs.			
N	**Instrument**		
24 & 25	Probed recall, RAN (color & pictures)		
Independent Variable		**Dependent Variable**	
Disability		Recall & RAN	

data that you enter and your imagination. For example, I recently required a substantial number of articles in a journal that was unavailable in our library. I used the form letter feature of the database to create an interlibrary request form that emulated the local library's form. This took approximately the same amount of time to produce as filling out one of the library's forms. I searched for all *Action* fields that contained the entry "order," and printed out all the interlibrary requests directly from the database.

CONCLUSION

Database management programs are flexible tools for anyone who must manipulate large amounts of data. While research notes can overwhelm manual data management tools, the computerized database can manage large data sets with ease. However, the process requires the investment of effort.

Computerized databases do not shorten the note-taking process. Entering the basic data, even when cutting and pasting from other databases or scanned text, is time consuming. Where the computerized database saves time is in the extraction of information. If you have 150 articles on phonological awareness, manually finding specific concepts or populations within that collection could require substantial amounts of time. The same process could be accomplished in seconds with a computerized database.

The computerized database is not the answer to all data management problems, but it can be a powerful tool. If you are willing to expend the effort to store your notes in a computerized database, the program can make organizing and retrieving that information much easier.

REFERENCES

American Psychological Association. (1994). *Publication manual of the American Psychological Association* (4th ed.). Washington, DC: Author.

American Psychological Association. (1997). *Thesaurus of psychological index terms* (8th ed.). Washington, DC: Author.

Apostle, L. (1987). Personal file management for the health sciences. Washington, DC: George Washington University. (ERIC Document Reproduction Service No. ED 281 560)

Approach Software Corporation. (1994). *Lotus approach user's guide* (Release 3.0). Redwood City, CA: Author.

Beckley, R. F., & Bleich, H. L. (1977). Paper chase: A computer-based reprint storage and retrieval system. *Computers and Biomedical Research, 10*(4), 423-430.

Bertrand, D., & Bader, C. R. (1980). Storage and retrieval of bibliographic references using a microprocessor system. *International Journal of Bio-Medical Computing, 11*, 285-293.

Brown, C. C. (1994). Creating automated bibliographies using Internet-accessible online library catalogs. *Database, 17*(1), 67-71.

Calvin, W. H. (1972). A computer-assisted personal literature reference system. *Computer Programs in Biomedicine, 2*, 291-296.

Chesley, R. (1989, February/March). Finding and using education research. *Thrust, 18*, 22-25

Gay, L. R. (1992). *Educational research: Competencies for analysis and application.* New York: Macmillan.

Hoke, F. (1994). Making the online connection with bibliographic-database software. *The Scientist, 8*(13). Retrieved January 8, 1999, from the World Wide Web: *http://www.the-scientist.library.upenn.edu/yr1994/june/tools_940627.html*

Holmes, T. F., & Gentry, C. G. (1979). A foolproof personal filing system. *Audiovisual Instruction, 24*(5), 40-42.

Houston, J. E. (Ed.). (1995). *Thesaurus of ERIC descriptors* (13th ed.). Phoenix, AZ: Oryx Press.

Jahoda, G. (1970). *Information storage and retrieval systems for individual researchers.* New York: Wiley-Interscience.

Jahoda, G., Hutchins, R. D., & Galford, R. R. (1966). Characteristics and use of personal indexes maintained by scientists and engineers in one university. *American Documentation*, 17(2), 71-75.

Lipsett, F. R., & Haycock, A. B.(1982). Preparation of personal bibliographies using a large computing facility. *Journal of Chemical Information and Computer Sciences, 22*, 1-4.

Ringel, M. (1980). Filed for easy reference. *Medical Group Management*, 27(1), 40-41.

Rowell, P. P., & Utterback, N. (1984). Scientific literature currency and organization using a microcomputer. *ONLINE, 8*(1), 18-21.

Singer, K. (1979). Where did I see that article? *JAMA, 241*(14), 1492-1493.

Soper, M. E. (1976). Characteristics and use of personal collections. *Library Quarterly, 46*(4), 397-415.

Stigleman, S. (1994). Bibliography formatting software: An updated buying guide for 1994. *Database, 17*(6), 53-65.

Stigleman, S. (1996). Bibliography programs do Windows. *Database, 19(2)*, 57-66. Retrieved January 13, 1999, from the World Wide Web: *http//web3.searchbank.com/infotrac*

Wanat, C. (1985). Management strategies for personal files: The Berkeley seminar. *Special Libraries, 76*(4), 253-260.

Wright, K. (1992). Bibliographic database management in an R&D organization: Information specialists helping end-users. *Database, 15*(3), 35-40.

Alex C. Pan

Using Technology to Promote Teacher Action Research

Infusing technology in teaching and learning has become a national norm. Presently, most classrooms have some computers, and in some schools, computers have become a required component in the curriculum. Computers are generally deemed to have potentials for enhancing teaching, learning, and teachers' productivity. Teachers generally are expected to have good computer skills and know how to integrate technology effectively into the curriculum. However, in reality, after advocating computer technology for over two decades, only a few schools have effectively integrated technology into their curricula, and still many teachers do not feel comfortable about using computers and even hold resistance or skepticism toward the usefulness of technological developments. Many researchers (Becker, 1994; Riel, 1994; Maddux, 1997) have discussed issues regarding computer uses in schools, such as factors that affect technology implementation, the frustrations that teachers are facing, differences between computer-using teachers and noncomputer-using teachers, and the gap between teachers in urban and rural districts. With many variables that challenge the school technology implementation, the consequences are less satisfactory than expected, and computers in the classroom were often left unused.

In fact, if implemented properly, technology has a great potential to

ALEX C. PAN is Assistant Professor, School of Education, The College of New Jersey, P.O. Box 7718, Ewing, NJ 08628-0718. E-mail: pan@tcnj.edu

[Haworth co-indexing entry note]: "Using Technology to Promote Teacher Action Research." Pan, Alex C. Co-published simultaneously in *Computers in the Schools* (The Haworth Press, Inc.) Vol. 15, No. 3/4, 1999, pp. 81-99; and: *Information Technology in Educational Research and Statistics* (ed: Leping Liu, D. LaMont Johnson, and Cleborne D. Maddux) The Haworth Press, Inc., 1999, pp. 81-99. Single or multiple copies of this article are available for a fee from The Haworth Document Delivery Service [1-800-342-9678, 9:00 a.m. - 5:00 p.m. (EST). E-mail address: getinfo@haworthpressinc.com].

help improve the way that teachers teach and students learn. To infuse computers in the school, teachers should not only possess good knowledge and skills about computers, but also know how to integrate technology into curricula and to engage their students in dynamic learning activities with computers. In other words, we need enthusiastic technology-literate teachers to teach with computers. To have computer technology make a profound impact on education, Schwartz and Beichner (1999) proposed a three-level Delta Principle, which contains (a) technology used as originally intended; (b) technology applied in new ways; and (c) application would not exist without the technology. The Delta principle actually advocates an implementable approach, with which technology will become an integral part of education, and both teachers and students will unconsciously depend on the power and convenience that technology can provide. Teachers especially will find great value in technology and so change their way of working.

One widely known phenomenon showing ineffective school practice of computer integration is the inadequate involvement of teachers in actively using and integrating technology into their curriculum. Struggling with insufficient time and meager resources, most teachers do not have a clue as to how to integrate computers in their own teaching. To address the issue of how teachers can accomplish technology integration, this paper proposes a field-tested plan that helped teachers learn to apply technology in their professional practice via classroom action research. Teachers' participation in action research of teaching and learning with computers is essential for technology restructuring in the school. Troen and Boles (1996) emphasize that teacher research is crucial to help teachers become the experts and transform schools. Moursund (1998) also states that, in essence, each teacher who is facilitating routine student use of information technology (IT) in the classroom is performing research experimenting with the effectiveness of such IT use. However, there is little research about how teachers' classroom research of IT has affected students' learning as well as teachers' teaching.

TECHNOLOGY-BASED ACTION RESEARCH FOR TEACHERS

When addressing the needs of technology integration in the public school, university teacher preparation programs naturally assumed leadership; however, many of the efforts failed to address the issues and were discredited. Many higher education experts took initiatives to conduct research exploring relevant issues and offered insights about which directions to go and what teachers should be doing about technology. Some university teacher educators provided training and technology-related courses to help teachers catch up with the new technology for education. However, efforts and initiatives like these would not work without modifications. In fact, teachers need more than just training or guidance to integrate technology in education. They need critical experience with which they can establish their own approaches of integrating technology in their own practice. Taking a reality check regarding how technology is used in the classroom, researchers such as Zehr (1997) and others indicate that most school teachers do not feel comfortable or ready when teaching with new technology. Many teachers, unfortunately, do not have sufficient access to technology resources, such as software and hardware, to engage their students in meaningful technology-enhanced activities. Although a small number of teachers are fortunate enough to get desired technological resources, they still don't know how to implement technology in instruction. They do not have a chance to construct the needed knowledge, experiences, and strategies for making good judgments about what to do and how to do it to inspire their students with technology.

Another factor that discredits the university teacher educators' efforts is the big gap between the advocacy of best practice from higher education and the real practice in public schools. It has been widely discussed among teachers that university professors have become ignorant of the classroom reality. Sengupta (1997) indicated that education professors are "out of sync" with public school teachers and some fundamental issues like what and how teachers should teach. In the application of educational technology, oftentimes teachers criticize university teacher educators who advocate unrealistic goals that are difficult to implement or unattainable. For example, Stephenson (1997) states that significant differences exist in the expectations about students' performance in computer education between schoolteachers and university professors. Such discrepancies make it hard to convince teachers to adopt the agendas and suggestions from higher education.

In spite of higher education efforts, engaging teachers in active computer-based action research is considered one of the most effective means of helping teachers to catch up with technology. Cochran-Smith and Lytle (1993) suggest that teachers should not just be passive followers, but behave like architects of study and generators of knowledge. Such a radical shift from passive information receivers to active action researchers transforms the notion of research on teaching and provides a radical challenge to assumptions about how teachers learn and what constitutes a knowledge base for teaching. If teachers started conducting computer-based research in their own classrooms, they would soon become sensitive to the real needs of technology and be anxious to become masters and generators of the new knowledge. Engaging teachers in active classroom technology-based research can make a significant impact on classroom practice. Teachers will not only learn how to use technology but also integrate technology in their professional practice where they also engage their students in active learning and research. The fact that computer technology has impacted the research process cannot be overestimated because computers have touched virtually every aspect of that research process (Shermis, Stermmer, Berger, & Anderson, 1991).

Teachers' action research on technology may also foster a new way of working at school. Sagor (1997) identified three major factors that may affect teachers' performance and lead to devastating consequences for educational change and reform: (a) work in isolation, (b) are generally excluded from knowledge production, and (c) are subjugated to external quality control. These factors can apply to the technology application in the school. Just as Sagor suggests, the collegial teacher research may improve the working situations at school. Teachers collaborate and cooperate by sharing experience and information with one another and by working together. If teachers become experienced in using computers for teaching, they are very likely to continue such a practice. If they have realized the power and the ease of use of technology, they will be happy to use it and apply it in other work. If teachers have conducted action-research of teaching with computers, they will become actively involved and will likely continue to grow with technology. In fact, how we help teachers understand the significance of the research and hold positive attitudes toward the research tasks is the key to inspire their enthusiasm and build their confidence in technology. The technology-based research by teachers is the best means to get teachers interested and involved. Teachers will appreciate the technology-based action research because it is closely related to their life and work.

THE TECHNOLOGY-BASED ACTION RESEARCH (TBAR) MODEL

The Need for a TBAR Model

As Kelley and others (1994) pointed out, teachers' computer needs have become difficult to address. Teachers in computer application classes constitute a widely mixed group with different interests, backgrounds, experiences, and computer skill levels. From my experience conducting computer courses, I find that no single fixed agenda can satisfy the needs of every teacher in the class. Usually the participating teachers prefer to learn computer applications with an easy-to-follow approach. However, it is difficult to provide an adequate pace for computer instruction to match teachers' diverse skill levels, interests, and expectations. For nearly every task, some teachers will find it difficult while others find it easy. The majority of teachers prefer to have more time to practice on their own but the fact that they are at different stages and directions makes it challenging to conduct the instruction effectively and efficiently. The most serious concern actually comes from the fact that most teachers are not able to apply what they have learned to their work after they take the computer class. Computer literacy is treated as a separate skill, totally unrelated to what teachers should do at work. To address all these issues and to provide a solution, the TBAR model is indispensable.

The TBAR Model

The TBAR Model was derived and modified from my prior experience of teaching and working with practicing teachers. This model was intended to develop technology-based research and to provide a possible solution to current school technology integration challenges. The TBAR model was implemented in two computer courses in a technology-enhanced instruction program. In the first course, 21 in-service teachers learned to master essential computer application skills and gained new insight about how computer technology can be used to enhance education as well as to support their teaching. At the same time teachers were also exchanging ideas. In the second course, these teachers assessed their own needs of using technology at work and identified issues and possible areas where they could improve with computers. After reviewing the literature, and deciding what they wanted to do, they designed a research plan, coordinated the activities, prepared necessary materials and instruments, and conducted an action research to find solutions to the perceived challenge with the new technology. At the end of the second course, almost every teacher became a master of technology in his/her own school and continued to implement technology in the classroom and to provide leadership to other people in the school.

The implementation of the TBAR model is based on the following considerations: (a) Reasonable nurturing experiences are essential to help teachers master needed skills; (b) action research must tie what teachers have learned in the course to their own classroom; (c) collaboration between university faculty and other teachers is necessary to help teachers break the isolation, connect with other school teachers, and become enthusiastic about technology; (d) the self-awareness process is important to help teachers identify problems in technology integration and find solutions; and (e) role-shifting plays a key role to change teachers from passive information receivers to active researchers.

The TBAR instruction uses Bloom's taxonomy framework (1956) for planning. The goal of technology learning was to enable teachers to apply their computer skills to the real classroom setting where technology is integrated into content teaching. Bloom's Taxonomy contains six levels of learning tasks that provide the foundation for learning in the TBAR model (as shown in Table 1).

Phases of the TBAR Model

The technology-based action research model consists of five phases that are extended across two courses as described below. These two courses carry the same mission but have a different focus (as shown in Table 2).

Course One–Computer Applications

Phase 1. Redefine computer literacy for individual teachers. Computer literacy may mean different things to different people. Teachers

TABLE 1
Bloom's Taxonomy and TBAR Tasks

Bloom's Taxonomy/Levels	Defined Tasks	TBAR Tasks
Knowledge	Rote memory of facts or definitions	Knowing about existing educational technology and tools
Comprehension	Ability to explain the knowledge in one's own words	Demonstrating how to use the tool by doing computer tasks and projects
Application	Knowing when, where, and how to use the knowledge and skills	Knowing when, where, and how to Use individual technology and tools for various aspects of work
Analysis	Taking an object or task apart to see how it is constructed	Understanding how technology tools can contribute to solutions to challenges at work
Synthesis	Combining parts of concepts to create a working whole	Drawing on computer resources to assist in problem solutions
Evaluation	Judging the relative merits of two or more ideas of systems	Fine-tuning the solutions to get the most desirable outcomes

with different content specialties and computer skill levels have different ideas of how technology may be used for their work. The selected components of computer literacy are based on individual teachers' personal experience and practical needs in using computers. Redefining computer literacy can help teachers build a high-confidence level for teaching and learning with technology.

Phase 2. Prepare teachers to know how to use technology. Teachers will master the basic computer application skills; that is, they will learn how to use computers to increase productivity as well as how to teach with computers.

Phase 3. Help teachers identify issues related to teaching and teacher's administrative duties. Organize discussions among teachers. At this stage, teachers will start brainstorming issues and topics for the action research. They will share ideas and discuss the status of technology use in their own school. They will also develop a working relationship among themselves so they do not feel isolated.

Course Two–Software Implementation

Phase 4. Help teachers refine their technological skills. Teachers will establish a well-planned research agenda with manageable topics. They will study the available examples, models, and other prior studies. They then design and develop necessary instruments and materials for their own needs. Most of the tasks they do at this stage are based on modifications of what they have accomplished in the previous course.

Phase 5. Conduct action-research, debrief, and share. Teachers will conduct research, analyze data, report the results, and present what they have accomplished in their projects and share the significant findings with other teachers.

The Scope of TBAR

Technology can be beneficial to education by fostering students' problem-solving and higher-order thinking, by enhancing student-directed learning, by providing diversity in instructional methods, and by becoming an effective management tool. To maximize the power of technology to serve education, numerous themes and issues are considered desirable for the TBAR projects. Teachers are highly recommended to research these areas:

TABLE 2
Focus of Two Courses: Computer Applications and Software Implementation

Computer Applications Course	Software Implementation Course
Learn about basic computer skills	Identify issues and problems
Learn to use computer as a tool in class settings	Modify computer skills, design and develop instruments for class
Use courseware for content teaching	Conduct action research in one's own class
Do integrative computer projects	Address needs and issues in one's class
Work and share with other teachers	Feel confident about technology integration
Achieve a high level of computer skills	Become leaders in the school

1. How can technology make a difference in teachers' work?
2. How should technology be integrated into curricula?
3. Under what circumstances can students benefit from using technology?
4. What factors may contribute to effective teaching and learning in the content areas?
5. How can technology increase collaboration and cooperation among students?
6. How do my students react toward the use of computer software in learning a specific topic, such as math, science, language arts, writing, social studies, and others?
7. How can I help some of my struggling students learn better and accomplish more with computers?
8. How can I use the spreadsheet to teach my students the concept of programming?
9. How can I enhance the instruction with only one computer in my classroom?
10. How much instruction about computer operation should be provided for my students to learn a subject area such as math or science?
11. How can I challenge my students to think and solve problems with computers?

Because technology can help connect teachers, students, parents, and beyond the school boundary, teachers may want to expand their research agenda to the relationship between school, parents, and community. Zehr (1997b) points out that technology is making it easier for parents, business people, and community members to get involved in education. At the same time, it is forcing some educators to rethink what they mean by "school." In addition, some broader issues such as communication, effective management, and multicultural awareness may deserve serious attention.

The Practice of TBAR

For two consecutive semesters, as a requirement for two computer courses, Microcomputer Applications for Educators and Software Implementation, 21 in-service elementary and middle school teachers learned basic educational technology, applied the skills in their specialty areas of teaching, and conducted action research in their schools.

The TBAR model serves as a means to help teachers grow and mature in learning to integrate technology in education. Participating teachers started with minimal computer knowledge and skills, but they expressed a variety of concerns about issues of ineffective computer integration in the school. To learn to cope with inadequate resources and other hassles encountered at school, these teachers were charged with a mission to identify the needs for improvement and to conduct action research in their professional practice to explore possible solutions to the challenges or problems. The procedures of the action research are shown in Table 3.

Research Areas Covered

Teachers were encouraged to select an area about which they were most concerned. Table 4 shows the frequencies of various areas that were chosen for the action research. The Internet was chosen most frequently, along with multimedia-based development and courseware integration. This result suggests the most popular trend of school

TABLE 3
Procedures of the Action Research

Procedures
1. Select a topic, a subject area, or a theme
2. Provide a rationale of needs
3. State issues and concerns
4. Review existing literature and research
5. Identify of support and assistance for research
6. Describe research subjects
7. Design, develop, and use instruments
8. Develop timeline for implementation
9. Identify expected outcomes
10. Report progress
11. Report of results and findings

TABLE 4
Areas Chosen by Teachers Engaged in Action Research

Research Areas	Frequency
Internet or Web-based	11
Multimedia-based development	3
Training-centered/staff development	1
Courseware integration	3
School technology planning	1
Team work and cooperative learning	1
One computer classroom	1

technology as well as teachers' perceptions of priority in technology integration.

Difficulties and Challenges Encountered

The two courses were slightly affected by the slow machines, software and hardware unreliability problems, and other technical difficulties. In addition, there were other challenges:

Teachers' low computer literacy. Teachers' computer skills were generally low to start with. Teachers' vision about educational technology was often hindered by the lack of technological skills. Because most teachers possessed very limited technological skills, they had to spend plenty of time working on various technology projects before they mastered the skills and became comfortable to move to the next phase. Not until they feel comfortable about using technology in their work will they seriously become involved in implementing technology in the curriculum. It is a serious challenge to help teachers' efficiently master needed computer literacy.

Platform differences. To help these teachers acquire needed computer skills, the first course used a series of computer tasks ranging from productivity application (such as word processing, spreadsheet, database, graphics, and communication) to authoring development and courseware evaluation. Most teachers preferred to learn about applications that they could practice at home or school and demanded to use software across computer platforms. Fortunately, there are software programs available for both PC and Mac platforms and the resemblance between these two platforms minimizes the users' effort of learning and transferring. However, such a demand incurred more effort in instruction.

Wide range of interests and skill levels. Teachers came to the computer class with a variety of interests and skill levels. Some had extensive prior computer experience, while others were novice users. In addition, teachers came with different backgrounds and content expertise. Their priorities and concerns about computer learning sometimes were in conflict. Some math teachers, for example, were not as enthusiastic about clip art as the art teachers. Some middle school math teachers were not particularly interested in software intended for early grade levels. To address the general as well as specific needs of every teacher in the class, this course required that all teachers go through the basic skill training and work on projects to improve their productivity. They then worked on their specific area of interest by working with software from the college software library and shareware from the Internet and other sources.

Security issues in the school. Most school district administrators were supportive of teachers' efforts in doing research projects in the school. However, some administrators were resistant about teachers' work involving the school name and relevant information. A few teachers created personal Web pages intended for instruction but were refused publication either on the school site or on the public site for security reasons. These frustrated teachers had no choice but to administer their action research using off-line browsers.

Scheduling problems. A few teachers complained that their computer lab was fully reserved by other groups, such as the writing classes, throughout the entire year. They were blocked out and were very frustrated. A special arrangement was made so that teachers could borrow computers for their classes, or they took their students to the university computer lab for their research.

Significant Results of TBAR Practice

The TBAR connected the technology learning to the actual needs in the school. Teachers learned to apply their computer skills to improve work productivity as well as teaching effectiveness via the teacher action research. It also transformed these teachers from passive information receivers to active participants or contributors of new information.

The TBAR experience changed the participating teachers' behavior in the school. Most of these teachers became comfortable and confident in using technology at work. Many began to take leadership in promoting technology in the school. After they took initiatives in the school to conduct research, they attracted attention from the administrators as well as their fellow teachers.

Most teachers become less threatened about the fact that their students outperformed them with technology. As described in Kusch, Pan, Bohm, and Stein (1997), some teachers developed a new relationship with their students, learning and exploring technology together, and found it mutually beneficial.

The TBAR helps participating teachers recognize the value of technology for instruction. As stated in Pan and Zbikowski (1997-1998), these teachers realized that computers were not just toys for making bells and whistles, but tools to generate meaningful pieces such as writing, presentations, and multimedia applications. The Internet does not simply provide tools for sending E-mail or browsing but can be used to produce dynamic learning moments to engage students in active tasks. The school administrators became aware of technology needs as teachers requested needed resources and assistance.

Teachers' Comments

Most teachers considered the action research a positive, meaningful approach to helping them actively learn about technology and software integration. Although most teachers appreciated educational technology and planned to adopt it in their own practice, still a few teachers were frustrated by the school reality. The following are some comments made by these teachers.

> Learning about technology and using technology did not mean much to me until I used it in my own class and found it well received by my students. I now depend on it at work.

> Some research studies suggest unrealistic things for teachers to do. They should really stand in our shoes before they tell us what's working or not.

> Making technology work in my school is certainly a challenge. Getting needed software, scheduling the computer lab, making sure everything is working, and other things took too much time. Using technology in my school is nothing but punishing.
>
> I took some computer workshops before but I wasn't able to apply the knowledge and skills to my work. After conducting the action research in my class, I now feel more comfortable using computers in my classroom.
>
> After I used computers in my class, my principal approached me and asked me to give a workshop for other teachers, sharing what I have accomplished. Several teachers in our building also came to me for assistance with their computer problems. I understand I didn't know too much more about computers than these other teachers, but I like the positive feedback that I received in my school.

ISSUES AND DISCUSSIONS

Many Teachers Feel Threatened by Being Outperformed by Their Students

Students generally seem more technologically interested and can pick up computer skills faster than adults. Many teachers feel threatened by their students' technology skills. It is important to help teachers overcome this anxiety and feel ready. Students' expertise should not be deemed as a threat to teachers. Teachers may turn skillful students into teaching assistants or role models for technology-enriched class activities.

Teachers Are Concerned About Insufficient Technological Resources

Although there are plenty of resources on the software market as well as on the Internet, most teachers find it difficult to locate good resources to meet instructional needs. In fact, there are many ways to acquire valuable instructional resources, such as finding reviews from educational journals or magazines, consulting with experienced users, checking with vendors, and requesting and testing sample programs from vendors.

How to Assess the Action Research

The TBAR model has provided a potential for engaging teachers in active learning and applying technology to their work with action research. The action research was assessed by various issues, including the following. How well are the instruments designed and developed to address the problems? Do research questions address the specified needs? Do the results of the action research reflect the real issues of using technology in the school or provide a solution to the challenge? The action research was also assessed based on a series of success indicators listed in Table 5.

University Faculty Role–Facilitation and Collaboration

The TBAR model has created an environment conducive to action research and technology learning. In the past, academic research and

TABLE 5
The Success Indicators

	Success Indicators
1.	The use of technology is an obvious component of the lesson or the research.
2.	Technology is used to solve real problems or to address real issues.
3.	Teachers are able to work out a plan to use technology in their workplace.
4.	Teachers are able to explain why technology is or is not used in a particular task.
5.	Teachers are able to voice their concerns and opinions for using technology with a global aspect.
6.	Teachers consider continued use of technology after they finish their coursework.
7.	Technology is used to impact student learning.
8.	Teachers can describe how technology is best used to address classroom teaching.
9.	Teachers can describe how a particular student is benefiting from technology.
10.	More exciting incident or results were happening than the teacher originally expected.
11.	Students are interested and enthusiastic about learning with computer.
12.	Teachers will have difficulty or feel uncomfortable if the technology piece is missing in their future work.

school reality were disconcerted. Most technology-based research was conducted in higher education, and in many cases, teachers were included only as subjects, but not as researchers. As a result, regardless of what researchers suggested to improve the classroom teaching with technology, teachers continued what they were currently practicing. This discrepancy should be reduced to the minimum so that the research findings can impact teachers' professional practice in the school. The TBAR model placed a major emphasis on teachers' initiatives. University faculty played a role to facilitate learning, to coordinate learning activities, and to collaborate with schoolteachers. The TBAR model also promotes a partnership between university faculty and schoolteachers, which in turn, promotes the partnership between teachers and students as teachers involve their students in the process of learning to use technology.

Desirable Action Research Projects

Any effort that may contribute to teachers' successful learning of technology applications and implementation is deemed an appropriate project. Some research topics presenting crucial focuses to reflect the true needs of using computers to change teachers' work and addressing issues in teaching and learning with computers are considered more important. Such topics include staff development issues, developing innovative strategies to motivate students, using computers to promote thinking and problem solving, and others. The process of engaging teachers in finding their own direction in technology integration is the most significant aspect of the TBAR model, which teachers will continue to use and grow with new technology in their professional practice.

Technology Promotes Research and Research Promotes Technology

The TBAR model promotes both research and technology. Participating teachers start by learning basic computer applications and apply learned skills to teacher administrative tasks and content teaching. After mastering essential technology skills, they conduct action research to address the needs of technology integration at work and to explore related issues regarding instructional effectiveness and teacher efficiency with computers. In spite of the hassles and inconveniences, teachers manage to conduct the research in their selected areas. Many of them feel more appreciative about what computers can offer after they find a way to implement technology in their classroom. In addition, teachers become leaders in their school, promoting technology integration. The relationship between technology and research is illustrated in Figure 1. Teachers' technology literacy helps smooth the process of maximizing the power of technology for action research. The action research helps teachers address the technology needs and explore effective ways of technology integration at school. The results of the action research help teachers attain a higher level of confidence and technology literacy.

Suggestions

Based on the results of the TBAR practice, several suggestions are made below to enhance teachers' technology literacy as well as to improve technology integration in the school.

1. University teacher education programs should provide a practical experience for teachers to master the essential computer skills required for their work in a more efficient manner.
2. Higher education faculty should also pay close attention to the reality issues in public education. Teacher education professors should provide necessary computer learning experience to teachers and help them apply the computer skills in their school daily life. Technology cannot be successfully integrated and thrive until teachers are convinced that technology is useful and easy to handle. Both university faculty and schoolteachers should examine the obstacles for technological implementation together.
3. Teachers should be encouraged to conduct action research projects in their professional areas and to involve school administrators in the process. Administrators should be encouraged to provide incentives and support for teachers to explore and find the best way to use technology at school.
4. University teacher educators should increase connection and communication by helping teachers collaborate with each other so that they can break the unnecessary isolation, share with, and learn from other teachers. The ties between university teacher education programs and public schools can be strengthened by establishing collaboration between the university faculty and the public schoolteachers.

FIGURE 1
Relationship between technology and research.

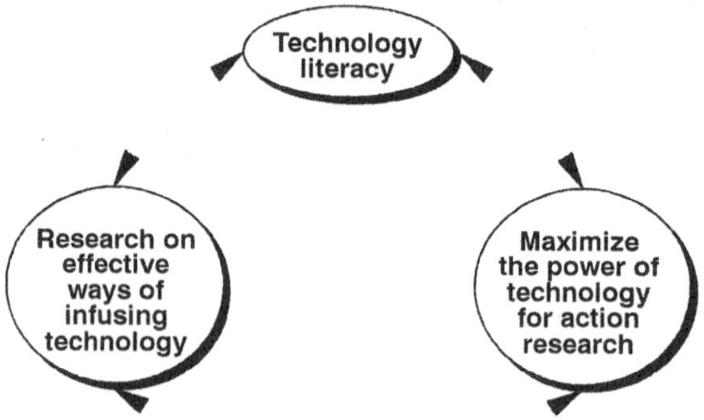

CONCLUSION

Technology integration in the curriculum has become a national norm. However, in reality, educational technology has been inadequately implemented in schools. Many teachers are struggling with meager resources and poor support for technology integration. As mentioned in this paper, teachers need more than just technology resources and instruction of what they should do. They need practical experiences of finding their own path of technology integration. To improve the current situation and to meet the challenge of infusing technology in the school, this paper proposes a field-tested TBAR model to engage teachers in active empirical research in their classroom, through which teachers master essential computer skills and learn to tackle the challenges they face by exploring innovative ways to implement technology in their schools. The TBAR model suggests that teachers learn to integrate technology best by applying what they have learned to their own work at school via the action research. To ensure the success of this model, more communication and collaboration between teachers and university faculty is indispensable.

REFERENCES

Becker, H.J. (1994). How exemplary computer-using teachers differ from other teachers: Implications for realizing the potential of computers in schools. *Journal of Research on Computing in Education 26*(3), 291-321.

Bloom, B., Engelhart, M., Furst, E., Hill, W., & Krathwohl, D. (1956). *Taxonomy of educational objectives: Handbook 1, cognitive domain.* New York: McKay.

Cochran-Smith, M., & Lytle, S.L. (1993). *Inside/Outside teacher research and knowledge.* New York, NY: Teachers College Press.

Kelley, A., Santiago, R., & Cooper, S. (1994). Is programming dead in teacher education? *Journal of Computing in Teacher Education, 11*(1), 19-22.

Kusch, J., Pan, C.A., Bohm, G., & Stein, H. (April 1997). Search for motivational software by teachers and students. *Technology and Teacher Education Annual,* pp. 31-34. Charlottesville, VA: AACE (Association for the Advancement of Computing in Education).

Maddux, C.D. (1997). The newest technology crisis: Teacher expertise and how to foster it. *Computers in the Schools, 13*(3/4), 5-12.

Moursund, D. (1998). Is information technology improving education? *Learning & Teaching with Technology, 26*(4), 4-5.

Pan, A.C., & Zbikowski, J.M. (winter, 1997-1998). Making meaning interactively with hypermedia–from bells and whistles to music: Using multimedia for literacy development. *WSRA Journal, 41*(3), 46-50.

Riel, M. (1994). Educational change in a technology-rich environment. *Journal of Research on Computing in Education, 26*(4), 452-474.

Sagor, R. (1997). Collaborative action research for educational change. In A. Hargreaves (Ed.) *Rethinking educational change with heart and mind,* pp. 169-191. Alexandria, VA: Association for Supervision and Curriculum Development (ASCD).

Schwartz, J.E., & Beichner, R.J. (1999). *Essentials of Educational Technology.* Needham Heights, MA: Allyn & Bacon.

Sengupta, S. (October 22, 1997). Are teachers of teachers out of touch? *New York Times.* p. B8.

Shermis, M.D., Stermmer, P.M., Berger, C.F., & Anderson, G.E. (1991). *Using microcomputers in social science research.* Needham Heights, MA: Allyn & Bacon.

Stephenson, C. (1997, January). Finding common ground between high schools and universities. *Journal of Computer Science Education, 11*(2), 6-7.

Troen, V., & Boles, K. (1996, August). Teachers, take charge! *Creative Classroom.* pp. 20-25.

Zehr, M.A. (1997, November 10). Teaching the teachers. *Education Week, 17*(11), 24-26.

Zehr, M.A. (1997b, November 10). Partnering with the public. *Education Week, 17*(11), 36-38.

Niki Davis
Roger Trend

Developing and Researching Information and Communication Technology as a Research Tool

Information and communications technologies (ICT) comprise a major focus within educational research. This paper discusses the need

NIKI DAVIS is Professor, Educational Telematics and Director, The Telematics Centre, School of Education, University of Exeter, Exeter, England, UK EX1 2LU. E-mail: n.e.davus@exeter.ac.uk
ROGER TREND is Senior Lecturer in Education, School of Education, University of Exeter, Exeter, England, UK EX1 2LU. E-mail: r.d.trend@exeter.ac.uk

ISDN case studies were within the project supported by the U.K. Department of Employment Learning Methods Branch and British Telecom PhC.

The environmental questionnaire is the intellectual property of Dr. Patrick Dillon and was developed within the project Teaching over SuperJANET and ISDN in Higher Education funded by the U.K. Joint Information Systems Committee New Technologies Initiative.

The authors acknowledge the support and work of all our colleagues in the Telematics Centre in the University of Exeter, especially Paul Thornton. The ICT Educational Research Centre in the U.K. Virtual Teacher Centre is supported by BECTa and is the creation of the first author. It is available at *http://www.ngfl.gov.uk/* in the Virtual Teacher Centre "Library and Research" section. You may also link to it from *http://telematics.ex.ac.uk/* ERC

The design of the ICT Educational Research Centre was informed by the design of the University of Exeter Virtual Study Centre, supported by funding from DG XIII_c Telematics Applications Program Project "Telematics for teacher training."

[Haworth co-indexing entry note]: "Developing and Researching Information and Communication Technology as a Research Tool." Davis, Niki, and Roger Trend. Co-published simultaneously in *Computers in the Schools* (The Haworth Press, Inc.) Vol. 15, No. 3/4, 1999, pp. 101-115; and: *Information Technology in Educational Research and Statistics* (ed: Leping Liu, D. LaMont Johnson, and Cleborne D. Maddux) The Haworth Press, Inc., 1999, pp. 101-115. Single or multiple copies of this article are available for a fee from The Haworth Document Delivery Service [1-800-342-9678, 9:00 a.m. - 5:00 p.m. (EST). E-mail address: getinfo@haworthpressinc.com].

to move to a more systemic approach to research in education. In order to justify and analyze this need for a strategic and coherent view across the various inter-related systems within education, the authors draw upon their experience of the Exeter "Development and Research" approach to the study of ICT in education. We start with the research that underpinned the first case studies that used ISDN technology in European education. We end with the 1999 launch of our ICT Educational Research Centre on the Internet within the United Kingdom Government's Virtual Teacher Centre.

There is an analogy with the convergence that has occurred in the technologies of computers, telecommunications and broadcast, and research to underpin the fundamental reappraisal of the application of technology across the various educational sectors including lifelong learning. This is not simply because research is bound to be influenced by the context and society within which it operates and where it must be applied, it is also due to the pervasive nature of technology.

The nature of information technology changes the way people and organizations work. The MIT'90s project categorized this change for commercial organizations into three stages: integration, evolution, and revolution. That final revolutionary stage is one in which technology forces the organization to re-evaluate its purposes along with its wider groupings, including both "market" and "supply" chains. We suggest that the increasing uptake of technology for research will also have a similar effect. This will force researchers to reappraise what they do, with whom and where. In educational terms, ICT is changing the ways (some) teachers and schools work, moving the teacher from center stage to the role of a facilitator or "guide on the side." We also expect to see many educational researchers move to the role of "guide on the side" as they empower the users of research, including teachers and policymakers, to use technology as a research tool to inform their own work.

An exemplification of this movement toward networking and facilitation that draws upon our research evidence is the focus of this paper.

THE EXETER APPROACH: DEVELOPMENT AND RESEARCH

Research undertaken by The Telematics Centre at the University of Exeter has often taken the approach of "development and research."

Note the reversal in the normal order of the term *research and development* because the development occurs first and then we research it. It is useful to distinguish the development and research approach from the "action research" approach. Action research also involves a cycle. However, the "action" is preceded by the researchers' evaluation of their own situation and informs their choice of the action to take. It often starts with the practitioner's evaluation of an educational issue, followed by planned action and then implementation. The cause-effect relationships are researched in a cyclical way. (A full account of the methods of action research in education can be found in Altrichter, Posch & Somekh, 1993.) The development and research approach, in contrast, takes as its starting point a development. In particular, we have taken the developments of multimedia resources and/or telecommunications. This is applied within the current educational system to the best of our ability. The development and research approach is not treated as any kind of unbiased experiment, but as the best we are able to do with a new resource with the aim of enhancing education, particularly teacher education. "Best" at this initial stage is often defined in practical terms. Although it may be informed by previous research, in practice there has been little to draw upon. Our aim is therefore clear, but the majority of the objectives are identified and modified as we progress in the application (and research) of the new resource.

Our strategy often has a twofold structure:

1. First, we explore the new application or resource, learning its strengths and weaknesses. These are discussed and reported on within the research group and less formally with colleagues from a range of subject disciplines within education.
2. In parallel, we identify people and contexts to which the resource may be valuable. When identifying these foci, we look particularly for opportunities and enthusiasm. We are well aware that education is a complex system, so we identify part(s) of it. This requires extensive observation and negotiation with teachers, their managers, support staff, and students. Increasingly, it also requires negotiation with service providers, national and local government agencies, and the wider community, including parents and employers.

The identification of context(s) for the application of new resource(s) frequently leads us back to more experimentation and trials of both the resource itself and of our own educational concepts and pedagogy. For example, when applying Desk Top Conferencing over point-to-point digital telephone lines (ISDN2) for the first time, we faced the loss of much of the non-verbal communication between teacher and learner in real-time rather than delayed communication (Davis & Wright, 1995). In order to study this, we recorded and analyzed video of a well-rehearsed ICT teaching approach involving one teacher in one classroom. Through analysis of the video of our own classroom-based teaching plus associated professional reflection, we were able to devise a way forward, which we then refined further in the light of experience. The research literature was also valuable in explaining some of the human-computer interface issues and laying down limits for acceptable delays of transmission.

This groundwork, exploring both the new resource and the educational environment in which it will be deployed, permits us to identify and develop limited trials of new forms of ICT in a way that does not risk the educational experience of teachers or learners. Each trial then becomes the documentation of a case study in which best practice evolves through a modified form of action research with the assistance of the participants in the various schools and universities.

The School of Education at Exeter University provides a very rich context for our research. Within this large university department we find a wide range of educational practices and subject disciplines, together with initial and continuing professional development of teachers for most phases of education. Our colleagues, students and researchers also provide prompt evaluation and access to appropriate educational strategies. The second author is a member of our research team from this wider institutional context.

The final phase of our research approach has frequently gone beyond dissemination into business planning. We are therefore familiar with a range of business strategies appropriate to the educational and training markets. Early work funded by the United Kingdom government encouraged us to create our first business model in the style of an agency, with support from the University's Centre for Management Studies (Dowie & Kimmance with Davis, 1995). Extension funding resulted in a cost/benefit analysis of ISDN in education and training. This was one of the first attempts to create an economic model of ISDN in educational contexts. That cost/benefit analysis involved the approach of a simple balance sheet.

A more thorough discussion of the relevance and importance of a systemic approach to research has been submitted to a journal of educational research by the first author. It develops the agricultural paradigm of educational research and illustrates this though discussions of part of the major evaluation of the United Kingdom superhighway initiative: the Bristol On-Line Educational Network (BEON) project. It also discusses the ethical issues relating to the development and research approach, which are particularly difficult when collaborating with commercial organizations.

THE RESEARCH ILLUSTRATED

To illustrate the development and research approach, three case studies that explore the ways in which the given technology can fulfill several roles will be considered:

1. A teaching medium as the focus of research
2. As a survey tool that is useful for both research and teaching
3. As a tool for disseminating educational research that will also become a tool for teaching educational research

The blurring of boundaries may be noted here between purposes and the roles of participants. This is a common outcome because the new technology permeates systems in which it is deployed.

Case Study 1: ISDN in Teaching at the Geography/Science Interface

Geography and science are well-established subjects in the United Kingdom secondary (11 to 18 years) curriculum and each has its distinctive set of core concepts. However, at the geography/science interface, the earth sciences (currently labeled "geoscience") have always been problematic at this level of education for a host of sociological and historical reasons (see, for example, Trend, 1996).

Teaching expertise in the earth sciences, especially geology, is not distributed evenly across all United Kingdom secondary schools. In some, it resides in the science department, in others the geography department, in others both, and still in others, neither. The emergence of the United Kingdom national curriculum (since 1990) has had no discernible effect on this irregular pattern. It was hypothesized that online teaching would provide an appropriate model for teaching geoscience, especially for those schools lacking the necessary teaching expertise or for those schools where further geoscience enrichment is considered appropriate for a small number of pupils. It was also recognized early in the project that the approach would lend itself to in-service teacher education and support in this problematic area of the curriculum.

Various basic geological concepts were selected for teaching to pupils in the pilot school by Desk Top Conferencing using ISDN2 digital phone lines. The university-based tutor and the school-based pupils had access to a hands-off telephone and to common computer screens. Control of the screen contents could be passed from one to the other by mutual agreement, but only the tutor had full knowledge of the image bank contents. No video link was involved.

Each of the 10 sessions was designed to involve one university-based tutor and two school-based pupils for about 40 minutes. Each session focused on basic concepts such as rock formation, fossilization, minerals, and plate tectonics. The overarching focus was the concept that lies at the heart of geology: geological time ("deep time"), and the subsequent research has confirmed the view that children's understanding of past "geo-events" is so unsophisticated that it is likely to inhibit further geoscience learning (Trend, 1998). For this reason, each session included items designed to engage children with the temporal aspects of the phenomenon in question. Prior to each session, a box of geological specimens was sent by surface mail to the school so that learning during the online session could be enriched and reinforced through pupil engagement with real geological materials. As appropriate, the tutor could clarify the link between screen, verbal stimuli, and the specimens. At the heart of each session was a bank of high-quality photographic and line-art images that could be drawn upon by tutor or pupils as required. The photographs included local landscapes and were stored on photo-CD. A whiteboard was also available as a screen.

As expected, evaluation revealed few difficulties with the technology and considerable benefits in terms of learner motivation and enthusiasm. Less expected was the enthusiasm of the teachers at the school who participated actively but necessarily somewhat peripherally. This aspect has developed further into a range of services for teachers in schools. In particular the value of images first noted with photo-CD was further developed into an artist in non-residence service (see the Telematics Centre Web site for further information).

Case Study 2: Web-Based Research/Survey

The Telematics Centre at the University of Exeter became well known for its innovative case studies with ISDN, not least through a video sent to all United Kingdom universities showing the case studies (it was not possible at that time to show video over the Internet). The good reception given to the video case studies led to the redevelopment and research of these early ISDN applications for the Internet. ISDN lines are dedicated communications channels where messages are not packaged together sharing the bandwidth, so it was fascinating, although not unexpected, to be brought up short by the distributed architecture of the Internet. The original (ISDN-orientated) approach had to be completely redeveloped and this research is discussed in Davis, Wright, Still, and Thornton (1997).

A particularly useful application included in the collection had been recreated from an amalgam of a "slide-tape" approach to resource-based learning and survey research. Dillon, Wright, Still, and Thornton (1997) describe this approach of conducting research over the Internet with an interactive, image-based instrument. The teaching focused on environmental education and teaching with associated students' knowledge of both environmental issues and their attitudes toward a variety of environmental contexts. The pilot Web-based questionnaire was of 28 students following primary (5 to 11 years) initial teacher training courses: 18 studying geography and 10 studying technology as their main curriculum subjects. The Web site was also used with geographers across Europe. The questionnaire was offered in two versions according to the respondent's information technology equipment. The completed questionnaire was E-mailed to the researcher. Questions focused on respondents' perceptions and values concerning environmental issues and environmental education. Several questions were constructed around color photographs depicting a range of features, respondents being required to interpret and evaluate with several further stimuli. Figure 1 shows a section of the questionnaire which is also available in the tools section of the Educational Research Centre described in case study 3 below.

FIGURE 1
Questionnaire for teaching and researching environmental education on-line.

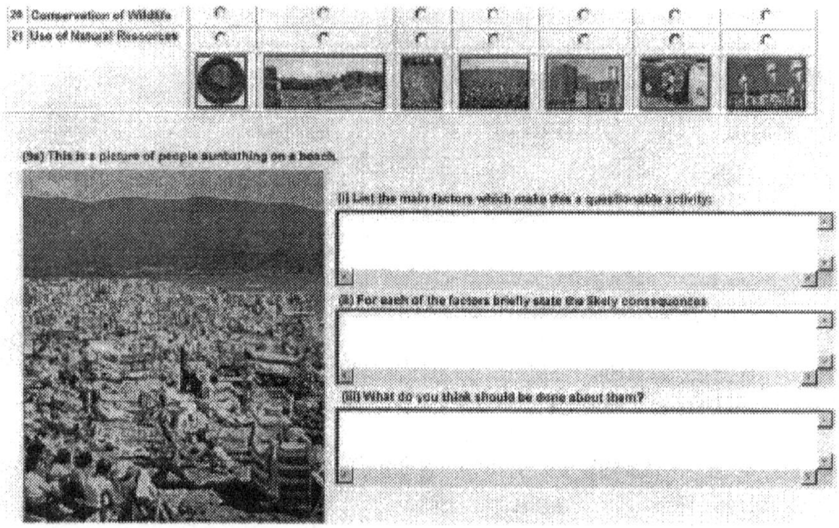

The benefits of the multimedia approach outlined in the preceding paragraph fall into several categories.

1. The technology allows rapid modification of the questionnaire, including text and images. Where copyright restrictions inhibit publication, an alternative replacement image can be inserted. Similarly, replacement images can be judiciously inserted to assess the instrument's reliability. Of course, the multimedia questionnaire can be developed very easily for different respondent categories.
2. The responses are obtained immediately and in a format that facilitates rapid processing of data that are used for both teaching and research.
3. The approach allows the possibility of respondents' selecting an image from an image bank.
4. The questionnaire can include links to other relevant Web pages (e.g., additional source material), a process equivalent to a second paper-based document in a traditional questionnaire.

5. Respondents can compare their input with that of other groups (the ongoing result of the research analysis). They may also be debriefed using ICT. In the trial the debriefing was provided by the researcher/teacher from another university through a speaker telephone and view of a Web page created by him.

Case Study 3: The Educational ICE Research Centre on the World Wide Web

The final illustration moves us to a research tool for dissemination that also is a tool for teaching educational research. This tool is itself currently in the first stages of development and research. Although expert researchers provide the "performances" as they carry out the teaching of educational research, the participants (teachers, researchers, and students) are actively involved and their contribution is also archived within the Web resource for the community of researchers and participants.

The U.K.'s National Grid for Learning (NGFL) offers significant potential for enhancing education and life-long learning. During 1998/99, a national curriculum for ICT is being implemented in initial teacher training. Trainee teachers, school-based teacher mentors, and others (e.g., librarians) entering the Web will therefore be keen to know about research evidence concerning the use of ICT for enhancing education and training. They will then apply this knowledge in their own work:

> [Teachers] "must demonstrate that they know (a) . . . about classroom-focused research and inspection evidence about the application of ICT to teaching their specialist subjects and where it can be found," and (b) "how to use ICT to join in professional discussions . . . including through the NGFL." (DFEE, 1998, p. 31)

The Telematics Centre in the School of Education, University of Exeter, is building an ICT Educational Research Centre, incorporating a seminar room and a library within the U.K. government's Virtual Teacher Centre (VTC), supported and coordinated with the British Education and Communications Technologies agency. The research center will be brought to life by "seminars" or interviews with leading educational researchers. Underpinning these performances will be the facilitation and growth of a community of researchers, many of whom are also practitioners (e.g., teachers). The program of events at the ICT Educational Research Centre will focus strongly on research outcomes of interest to schools relating to ICT in education. Sessions are being archived in hypermedia so that they remain available and can be enhanced over time by incorporating new research and resources as they become available. This will ensure a healthy balance between recent research and the archived activities and resources (including an archive of data and research tools).

It is proposed to progressively expand the participating individuals and groups to encourage them to take a central role in the ICT Educational Research Centre perhaps by providing a lecture or interview on an aspect of ICT research that draws upon their own work, or to undertake a review across a theme. A regular program of multimedia events is advertised. Each event will be linked to a time-limited "audience participation," which will be enhanced before being archived. Occasionally, adopting an approach used by broadcasting companies, a performance will be captured at a live event or through a videoconference. As appropriate, the experts will be assisted by the Telematics Centre's research and design team to create suitable multimedia "performances" coached in appropriate audience participation strategies. Much of this support will be provided using communication technologies themselves, including Internet and point-to-point videoconferencing. Academic rigor will be underpinned by strategies evolved from those for academic journals and external examination procedures.

The interface aims to invoke appropriate behavior from participants, such as those shown in the pictorial menu in Figure 2. An alternative mode of use draws upon the analogy of desk-based research, shown in Figure 3. Pictures and icons provide the menus. For example, (a) to hear a lecture the user clicks on the speaker, (b) to search a Web site the user clicks on a computer, and (c) to ask a question of the expert the user clicks on a phone (also to add new questions or a comment).

Resources will be built up in the ICT Educational Research Centre session by session. Research into Web sites is starting to demonstrate the need for sites to evolve organically, developing and maturing to suit the needs of providers, browsers, and active participants. Such evidence is emerging from the Web Evaluation (WEVAL) framework developed using the development and research approach in Exeter by Roger Poland and Paul Thornton (1998). Web sites that develop by involving providers and users across several sectors, especially those with complementary needs and resources, are likely to be the most

FIGURE 2
Pictorial menu of the ICT Educational Research Center (Embryonic version).

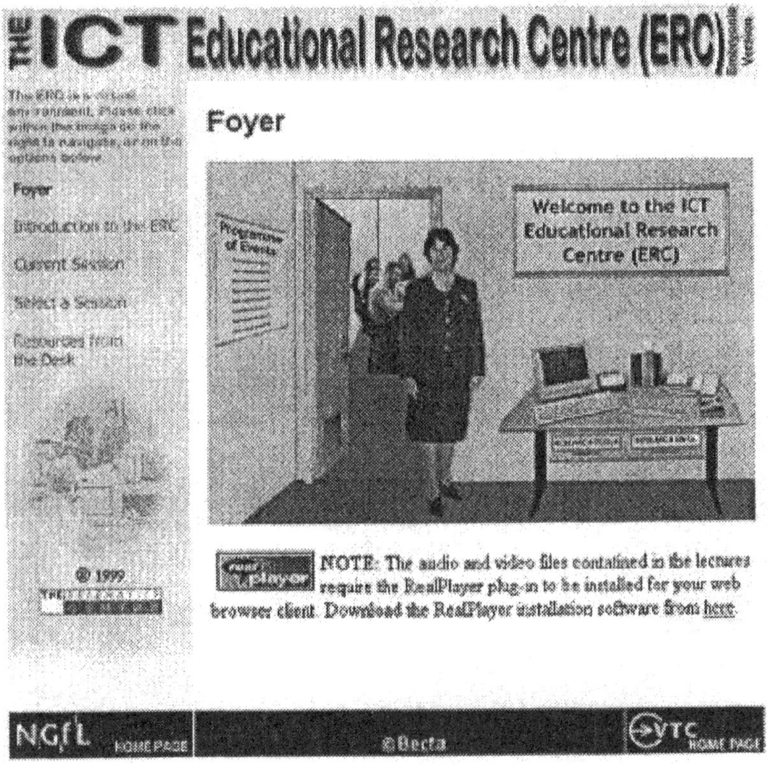

robust over time. This development and research approach to educational research, which underpins such Web site development, includes the active networking with many levels of a research community (including its users), stimulating and supporting them to collaborate in accordance with their needs.

Our current template for presentation by speakers has the following features:

1. Audio streaming of the lecture alongside its PowerPoint presentation.
2. A framed Web page so that each overhead transparency and its notes and other visual aids can be presented.
3. A "chunking" of the lecture into distinct elements so that users can easily listen to a chosen section and engage more deeply with the

discussion alongside it. This requires both capture of the lecture, and the creation, editing, and refinement of multimedia elements.

Our expertise and presentation style will grow as we practice this and, of course, we will be researching it with active participation of our users.

CONCLUSION AND A WAY FORWARD

The development and research approach described in this paper lends itself to systemic inquiry into the pedagogical issues associated with rapidly changing technologies. In all three case studies presented, the focus comprises a particular resource that has emerged essentially as a result of market forces and technological advance-plus-innovation. The resources have not emerged in response to perceived educa-

FIGURE 3
Desk-based research in the ICT Educational Research Center (Embryonic version).

tional needs. However, it is essential that this role be researched in a context of teaching, learning, and research. The familiar empirical research approach is inappropriate for such a scenario.

At the heart of the development and research model is the idea that the new "development" represents a resource that is currently available to teachers and learners but which,

1. is set to evolve rapidly or to be replaced by a similar resource which itself will evolve rapidly;
2. is open to a range of uses by teachers, learners, policymakers and researchers;
3. lacks a well-established protocol for its effective use; and
4. is likely to involve either evolution or revolution in some aspects of users' relationships.

These four characteristics provide various challenges for educational researchers. For example, what is the most appropriate research model for such a situation? To what extent are all school-focused research approaches destined to become modified by the expansion of ICT in schools?

The authors anticipate that the development and research model discussed in this paper will itself evolve and become more sophisticated with time, especially in light of its wider replication across the world. The basic D and R sequence is one which has already paid significant dividends, not only in the projects reported above but also in further projects reported elsewhere (e.g., Davis, 1999).

Undertaking development and research within the university context leads us, as educational researchers, to reflect more profoundly on the nature and strategies of educational research. This distinctive approach will encourage us and our collaborators to re-evaluate our practice. This parallels the challenge that information and communications technology sets teachers, one that can only be faced through re-evaluation of their practice of teaching and supporting the learning process (see, for example, the underlying emphasis on changes in the teachers' role in Somekh and Davis, 1997). Potentially, information technology applications can change the most fundamental aspects of teachers' practice: the process of communication, the people who undertake the teaching; collaboration, and professional development. For educational research, the issue is also one of communication and the identity of "the researcher." Can educational researchers stand back to guide new practice and thus empower all participants to research and develop their own practice? This paper suggests that such an approach is both necessary and valuable and that researchers will need to facilitate and guide users in the deployment of such research tools.

A potential revolution is underway in education and the situation may well be nearer to its critical mass than is widely appreciated. Many educational leaders, government policymakers and commercial organizations are already convinced that ICT will provide enormous benefits to learners in terms of both quality and efficiency. Such a (relatively) rapid revolution is even more likely, given (a) the U.K. government's current initiative to create a National Grid for Learning, (b) the related teacher training funded by the U.K. lottery, and (c) the European Commission's recognition that teacher training is a key issue for Framework V program of research and development, which was published in 1999.

In North America there are also many initiatives to foster appropriate use of ICT and the U.S. President's Commission provided guidance on the required research and development. Commercial organizations across the world also appear to predict a valuable and rapidly expanding new market for their services.

Educational research needs to adapt and promote educational development and research in order to inform this revolution. This paper has attempted to provide some pointers to a possible way forward. It has also discussed ways in which ICT may be deployed to assist educational development and research, in that order. The most important and fundamental challenge is for educationalists to achieve a refined understanding of the aims and process of education systems in the context of an ICT-literate society (noting, of course, the exclusion of enormous sectors of humankind from such participation). There is no time to lose: The revolution will not wait for educational researchers. Perhaps this revolution in education will provoke a revolution in educational research, empowering all participants to use and learn from its processes and findings.

REFERENCES

Altrichter, H., Posch, P., & Somekh, B. (1993). *Teachers investigate their work: An introduction to action research.* London and New York: Routledge.

Davis, N.E. (1993). *Cost-benefit analysis: ISDN in education and training.* (Report for U.K. Department of Employment). Exeter: Exeter University.

Davis, N.E. (1999). *A systemic approach to research in education, prompted by information and communications technologies*. Unpublished manuscript. Exeter: Exeter University.

Davis, N.E., Wright, B., Still, M., & Thornton, P. (1997). Pedagogy and protocols for collaborative teaching and research through SuperJANET and ISDN in higher education. *Innovations in Education and Training International*, *34*(4), 299-306.

Davis, N.E., & Wright, B. (1995). *Final report of the project: Multimedia communication services*. Exeter: University of Exeter.

Davis, N.E., Wright, B., & Still, M. (1993). *Final report and video of ISDN2 and computer networks to enhance initial teacher training*. Sheffield: Department of Employment.

Department for Education and Employment (1998). *Teaching: High Status, High Standards, Circular 4/98, Annex B: Initial Teacher Training National Curriculum for the use of Information and Communications Technology in subject teaching*. London: Author.

Dillon, P., Wright, B., Still, M., & Thornton, P. (1997). Conducting research over the Internet: An interactive, image-based instrument for investigations in environmental education. *Journal of Information Technology for Teacher Education*, *6*(2), 147-157.

Dowie, C.A.I., & Kimmance, J.K., with Davis N.E. (1995) *Business plan: Multimedia telecommunications agency*. Exeter: University of Exeter School of Education.

Poland, R. & Thornton P.C. (1998). *Web evaluation for education*. Presentation to Exeter University Research Student Conference, Exeter, February 1998.

Somekh, B. & Davis, N.E. (Eds.). (1997). *Using IT effectively in teaching and learning: Studies in pre-service and in-service teacher education*. London and New York: Routledge.

Trend, R.D. (1996). Fostering collaboration as a means of enhancing geoscience teaching in Devon, England. In D.A.V. Stow & G.J.H. McCall (Eds.), *Geoscience education and training in schools and universities, for industry and public awareness* (pp. 367-378). Joint Special Publication of the Commission on Geoscience Education and Training of the International Union of Geological Sciences and the Association of Geoscientists for International Development, AGID Special Publication Series, No. 19. Rotterdam: A. A. Balkema.

Trend, R.D. (1998). An investigation into understanding of geological time among 10-and 11-year old children. *International Journal of Science Education*, *20*(8), 973-988.

Wright, B. (Ed.) (1995). *Multimedia communications brokerage*. Sheffield: Department for Education and Employment.

Wright, B., Davis, N.E., & Still, M. (1996). *Selection of resources and technical issues relating to desktop and video conferencing*. Exeter: University of Exeter Multimedia Communications Services.

Jerry Willis
Muktha Jost

Software for Analyzing Qualitative Data

Over the past decade a number of specialized computer programs have been developed to support the analysis of qualitative data. Most of these programs, with names like Ethnograph and NUD*IST, were developed by qualitative researchers to support their own work. These early versions were sometimes informally shared with colleagues interested in doing the same types of research. Then, as the programs matured and interest in them increased, a number of the programs were picked up by commercial publishers or organizations for distribution. They can now be purchased by anyone. In this article we will review in some detail three of the most popular qualitative data analysis software packages currently available. For those who have decided to explore the use of qualitative software, we hope this article will help make tentative decisions about which programs are most appropriate for research. Before looking at programs, however, two issues need to be dealt with: the purpose of research and the issue of holistic versus atomistic approaches to data analysis. Both influence the type of analysis we consider appropriate and thus the type of software we prefer.

JERRY WILLIS is Professor, Curriculum and Technology, Center for Teaching, Learning, and Technology, College of Education, Iowa State University, Ames, IA 50010. E-mail: jerryw@iastate.edu
MUKTHA JOST is a PhD candidate, Curriculum and Technology, Center for Teaching, Learning, and Technology, College of Education, Iowa State University, Ames, IA 50010. E-mail: Mjost@iastate.edu

[Haworth co-indexing entry note]: "Software for Analyzing Qualitative Data." Willis, Jerry, and Muktha Jost. Co-published simultaneously in *Computers in the Schools* (The Haworth Press, Inc.) Vol. 15, No. 3/4, 1999, pp. 117-150; and: *Information Technology in Educational Research and Statistics* (ed: Leping Liu, D. LaMont Johnson, and Cleborne D. Maddux) The Haworth Press, Inc., 1999, pp. 117-150. Single or multiple copies of this article are available for a fee from The Haworth Document Delivery Service [1-800-342-9678, 9:00 a.m. - 5:00 p.m. (EST). E-mail address: getinfo@haworthpressinc.com].

THE PURPOSE OF RESEARCH

There are many traditions in qualitative data analysis, but most of them will fit into one of two broad categories: holistic or atomistic. These two traditions differ in many ways, including the underlying philosophies of science that support them. An issue that encapsulates many of the differences between these two approaches is the question of what we expect to gain from research. There are four basic answers to this question. Each is based on a general philosophy of science and has many implications for both how the research is conducted as well as how data is analyzed and written up.

General Theory

The purpose of postpositivist quantitative research is to discover general laws or law-like relationships. You do a study to test a theory. The local context is not important except for validity and reliability issues because the reason we do research is to find truths that can be generalized. Qualitative research based on this approach (Miles & Huberman, 1994a, b; Smith, 1993) is done for the same purpose as traditional quantitative research. Approaches like grounded theory (Strauss & Corbin, 1997) also generally have this as a purpose.

Local Theory

Some qualitative researchers acknowledge that a single study in a specific setting cannot produce Truth that can be generalized directly to other settings. Proponents of this approach attempt, instead, to develop a full understanding of the local context. They search for truth about the local context. Collins (1998) pointed out that it is common practice to use interview data to discover *facts* about a situation in an effort to build local theory (or general theory). The purpose of this form of interviewing is to mine the sources of data in order to understand what is really happening and then to communicate the essence of this to others. Other than accepting the limited generalizability of the results, local theory studies follow essentially the same guidelines as general theory research. Collins (1998), who advocates thinking of qualitative interviewing as a way of understanding a story you are going to tell, believes it is impossible even to get at local theory because interviewing is not an objective process, it is a very subjective and complex act of human interaction.

Objective Description

General and local theory research involves considerable inference. You develop generalizations from the raw data and emphasize those generalizations in the write-up. A third approach, also based on the assumption that qualitative research can lead to Truth about a situation, focuses on accurate description of the situation under study. Description is at the basis of both general and local theory research, but it can be a goal in itself. From the point of view of interpretive and critical researchers, trying to stick to simple, "objective" descriptions does not work. Description is not, in the words of Sergeant Friday of the old *Dragnet* television program, "Just the facts, ma'am." Description is selective; it is the interpretation of the describer, and therefore it is subjective.

Hermeneutic (Verstehen) Understanding

The purpose of descriptive and theory-based research is to identify the truth of the matter, or at least get as close as possible to it. It has been posited that Truth exists separately from the people who are studied, and it is separate from the researcher. While this may be a defensible position in the physical sciences, it is less likely when humans are studied because humans are reactive–their perspectives and views influence what they do, and their experiences influence their perspectives. The purpose of hermeneutic research, or to use a slightly more general term in German, *verstehen* is to understand the perspectives of humans. And, because understanding is situational, hermeneutic research tends to look at the details of the situation in which the study occurred. The result is generally "rich" data reports that include multiple perspectives. There can also be efforts to develop "theory" and test hypotheses, but as Kelle (1997) points out, the theories and hypotheses of hermeneutic research are not the same as the theories of descriptive and theory-based research. Hermeneutic theories are more like fuzzy, tentative explanations that ebb and flow, merge and diverge as research progresses. To develop and test theories in the tradition of the scientific method used by quantitative researchers, the theories and hypotheses must be precisely defined prior to conducting the research and linked to well-defined, observable behavior. This is not common in holistic qualitative research. The hypotheses and theories qualitative researchers typically work with are "(sometimes very vague) assumptions and conjectures about possible relations between certain domains. To examine these hypotheses means to return to the material in order to explore this possible relationship by a thorough analysis of textual data. This interpretive analysis of text (segments) may then form the basis for the clarification and modification of the researchers' initial (general or vague) assumptions. The notion of hypothesis testing would be rather misleading here, if one understands it as an attempt to falsify an empirically contentful statement" (Kelle, 1997, Online).

When verstehen is the purpose of research the researcher may work with fuzzy, incomplete, and emerging ideas that are called theories or hypotheses. They are quite different, however, from the theories and hypotheses researchers use when they are seeking Truth (general theory, local theory).

Often the result of hermeneutic research is not anything close to a theory. Often the result is lists of *exemplars* rather than generalizable truths (Smith, 1993). Postpositivists are generally uncomfortable with the idea of leaving lists of exemplars as the final product of research; they would like to transform them into theories that represent universals. "Interpretivists, of course, will have none of this at any level. They agree with Feyerabend (1981) when he says that, for certain traditions . . ., 'a list is not a mistaken first step on the way to a more appropriate definition: it is the only adequate form of knowledge.' . . . One learns to use a list, and thereby becomes part of a practical tradition, by playing around with or applying the list in interaction with others" (Smith, 1993, p. 141). Lists, however, are not simply different forms of postpositivist universals. "Items on a list are not rules that foreclose or predetermine the kind of judgments people make; they are values that influence those judgments" (Smith, 1993, p. 141).

Storytelling

Collins' (1998) work is an example of storytelling. He insists that meaning exists only in context and that qualitative interviewing does not uncover pre-existing truths. Instead it is a social process that constructs meaning through interaction. This approach does not attempt to present *the truth of the matter* as description does. Instead, it is an effort to represent in considerable detail the perspectives of participants in the process or setting being studied. While description produces one truth about the topic of study, storytelling may generate multiple perspectives as well as interpretations and analyses by the researcher and participants. Storytelling has much in common with hermeneutic research, but it puts more emphasis on representing the perspectives of the participants within a context of details about the setting or situation.

The purposes of research point toward certain types of data analysis. If you want to build a general or local theory (or an accurate description) that explains the truth of what you are studying, you may want to consider emphasizing *atomistic* approaches to data analysis. Atomistic approaches generally involve breaking the data down into segments, adding codes to the data segments, and then looking at relationships between the codes.

On the other hand, if you take a hermeneutic (interpretive) or narrative (storytelling) approach, some of the more holistic methods may be most appropriate. Holistic approaches tend to leave the data intact and to emphasize that meaning must be derived for a contextual reading of the data rather than the extraction of data segments for detailed analysis.

While we have drawn the distinction between atomistic and holistic analysis sharply, in practice many researchers use both forms of data analysis in their work. They use both types of analysis routinely for everything from general theory building to storytelling. In the following sections we will look at a number of the popular qualitative data analysis software packages currently on the market. Most of these programs support an atomistic approach to data analysis, as we have defined it here, but keep in mind that these programs can be used in many different ways, including to support hermeneutic and narrative research.

THE WHOLE VERSUS PARTS DIVIDE IN QUALITATIVE RESEARCH

If we crudely divide qualitative research into two traditions–a holistic and an atomistic approach–the differences in data analysis approaches become more obvious. Holistic researchers worry about the ability of any qualitative data analysis software to make code-and-retrieve data analysis both easy and sophisticated. They fear it will lead novices to use that approach even when it is not appropriate. And, since this group less frequently sees code-and-retrieve as appropriate, they do not find programs that help you do that very appealingly. The same is true of theory building. Since some researchers who take a holistic, in-context approach are not trying to build theories or laws that can be applied outside the situation studied, there is little reason to learn to use software that automates the process of building theories and testing hypotheses.

Naturally, atomistic qualitative researchers tend to see things from the opposite perspective. Support for storytelling is not very useful to this group because the purpose of their research is not to tell a story. It is to present useful generalizations and to convince people that the generalizations are accurate. This group will prize software that supports code-and-retrieve data analysis and theory building. Holistic researchers are more likely to prize software that supports understanding and storytelling.

Collins' (1998) work, which was discussed earlier, is an example of a holistic approach to qualitative data analysis. He worked with the transcripts of his interviews without separating significant segments from the surrounding context. Collins did not code his data. His approach is in marked contrast to that of Mulder (1994). While both researchers were working with interview data, Mulder did code her data, and she separated significant segments from the surrounding text. She actually created a new context by organizing the data into patterns based on perceived commonalities.

Holistic versus atomistic (coding) is an important distinction, but qualitative researchers take positions all along the continuum between emphasizing holistic analysis almost exclusively to considering coding and breaking the data down for analysis virtually a necessity. Lonkila (1995) expressed the moderate views of many who work toward the atomistic end of the continuum, when he said, "At least some kind of coding is needed in most qualitative research, but it is also possible that coding is overemphasized, given the fact that a large part of the qualitative researcher's work consists of interpretation and a fine-grained hermeneutic analysis" (p. 49). Some would take issue with Lonkila's assertion that most qualitative research calls for coding, however. We will not even try to offer suggestions about whether holistic or atomistic approaches are "better." They are both widely used in qualitative research. There are, however, differences in the types of software that researchers from these two traditions prefer and differences in the way researchers from these two traditions would use a particular piece of software. In the next section, we will review the three most popular qualitative data analysis packages currently available and look at the ways these programs might be used by both holistic and atomistic researchers. (Note: sources of information about all the programs mentioned in this paper are listed after the references.)

THE ETHNOGRAPH 5.0: A PIONEERING PROGRAM

The great majority of qualitative studies involve the collection of a large amount of textual data. That data are then subjected to an analysis that involves creating a set of codes and then "coding the data" so that text segments (words, sentences, paragraphs, comments) have one or more codes attached to them. Those codes indicate that the segment is relevant to a particular topic such as "feelings about school" or "plans after treatment." Before computers, coding was often followed by a cut-and-paste phase. The data, which was on sheets of paper, was cut apart. Each slip contained a coded text segment. All the segments with a particular code were put together. Often they were pasted to a large piece of paper or poster board. Then, with all the text segments with a particular code in one place, the researcher began a process of trying to derive some meaning from the coded comments.

Qualitative research is labor intensive, and a major reason the Ethnograph was developed was to reduce the amount of work. The Ethnograph is primarily a labor-saving tool. It helps you do the type of code-and-retrieve data analysis that you would have done anyway. We say "primarily" because there are some things you can do with Ethnograph that you would probably not do if you were working manually.

To illustrate how Ethnograph works, we have divided a typical code and retrieve study into seven phases. Please keep in mind, however, that while we are presenting these "phases" as separate, sequenced activities, the way they actually happen is not so neat and tidy. All phases can happen at the same time, and the work in many phases may be done over and over as the research progresses. In addition, the phases are not quite what we would propose if we were developing a set of phases for interpretivist or verstehen research. These are a bit more linear in feel and more focused on finding "facts" than we would propose. However, the seven phases do fit the process of qualitative research that programs like Ethnograph were designed to support.

Conception and Planning for the Study. In this phase you decide what you are going to study, and you work out the preliminaries such as gaining entry to the research environment. The Ethnograph plays no role in this phase.

Data Collection. In this phase you collect your data. We will assume that the research produces a body of text data such as interview transcripts or detailed field notes. (We do realize that qualitative research today often uses video, audio, and images as data. Only one of the programs, Atlas/ti, has strong support for this type of data.) The Ethnograph typically plays no role here.

Conversion to Data Files. Once you have your data in some form, such as a set of handwritten field notes, you must convert the data to one or more computer files. If your data are already in a word processor file, Ethnograph will let you import that file. You first create a "project" and give it a name such as "Community Leaders," and then begin to import the word-processing data files associated with the project. You might, for example, have a set of files that contain the transcripts of interviews with school administrators and another set of files of interviews with teachers who work for the administrators. Each file can be imported into Ethnograph and linked to the Leaders project.

You can also enter the data directly into an Ethnograph file. However, Ethnograph's editor is not very powerful. If you have a favorite word processor, you may want to use it to create the original version of the data file and then import it to Ethnograph. This approach may be quicker than using the built-in editor.

The term *import* is actually overkill for how the Ethnograph handles word-processing files. In reality it uses the cut-and-paste functions of your computer's operating system. To get a word-processing file into Ethnograph, you load the file into your word processor, use the Select ALL command to highlight the entire file, and then click the Copy command. That puts the entire file in the clipboard memory of your computer. Then you switch to Ethnograph, create a new data file, and use the Paste command. This works reasonably well, but keep in mind that none of the fancy formatting features of your word processor will be transferred over to the Ethnograph file. Also, the word-processing file may give Ethnograph some problems because it wants its files to be in a certain, precise format. Ethnograph has a Reformat command with several options to help you clean up the word-processing file and make it digestible.

Regardless of how you do it, you will need to learn the formatting rules for Ethnograph data files. The small data file presented here shows how a typical one looks. It was taken from the demonstration files that come with the program:

> +Focus Group-3M PHYS,1M ENG, 1F GEOL
>
> Note: Throughout the transcript, the female geologist is identified by W; the four males as M1, M2, M3, M4; the interviewers as B and J.
>
> J: Is there still a "good old boy" attitude?
>
> M1-4: Uh-huh.
>
> W: There really is. If you look in the engineering building you'll find that there's very few women who dare step through those doors and want to stay there.
>
> M2: But you'll find that there are a lot of people who come into college . . . I didn't know what I wanted to do, but I think that a lot of people go into college and they try out the classes here and there to find out where they might fit, and to find out what their majors are gonna be. If you go in and take an engineering course or something and find that there's not a welcome attitude there, you're not gonna go in that direction.
>
> W: These fields are really very sexist. No question.
>
> M3: Well, there is this attitude in general of the faculty, uhm, it's like they try to convince you this is very, very hard stuff and in effect that means, "Oh, I understand it, so therefore I'm quite the genius. If you can do it, then you must be too. But if you can't, boy, you've got nothing going for you."
>
> M1: And if you survive–not WHEN you survive–you only will have a basic understanding of a few things. It's like this kind of, uhm, like you're getting these conflicting . . .

The plus sign at the beginning of the first line indicates this is a heading with information about the sources of the data. Anything at the beginning of a line that is followed by a colon tells you who is speaking. Ethnograph is also particular about paragraph codes. You should not, for example, end each line of dialog by pressing the Enter key since that will put an end-of-paragraph code in the file at that point. Instead, let the text "wrap around" and begin a new line.

Code the Data. Once your data is in an Ethnograph data file, your next step is to code it. Coding involves creating a set of descriptors that apply to your data. Several authors have created lists of different types of codes:

> Miles and Huberman, 1994a: descriptive, interpretive, and explanatory codes
>
> Bogdan and Biklen, 1992: settings/context codes, definition of situation codes, perspectives held by subjects codes, subjects' ways of thinking about people and objects codes, process codes, activity codes, event codes, strategy codes, relationship and social structure codes, methods codes, and coding systems developed by others

These categories may or may not be useful. The important thing is to begin to develop codes that make sense given the data and the reason for doing the research. Sometimes, because your research focus has been the subject of earlier important studies, you may want to adopt a coding scheme that already exists. This approach is encouraged by some and discouraged by others. Kuczynski (1998) talks about the need for codes to develop from the data. He quotes Bakeman and Gottman (1986), two internationally respected qualitative researchers, who argue against borrowing codes. ". . . sometimes we hear people ask: Do you have a coding scheme I can borrow? This seems to us a little like wearing someone else's underwear. Developing a coding scheme is very much a theoretical act, one that should begin in the privacy of one's own study, and the coding scheme represents a hypothesis, even if it is rarely treated as such" (p. 19).

On the other hand, the manual for one qualitative software package, AQUAD (*http://www.aquad.de/index.html*) encourages the use of "pre-determined category systems." Adopting a coding system developed and used by someone else may be the easiest path to take for beginning researchers. A coding system can always be modified and revised as you analyze your data. A second approach, creating a basic coding system and then expanding and revising it as you analyze your data, is another common option.

Once the codes have been at least initially selected, the next step is to decide on your unit of analysis. A unit of analysis is the smallest segment of text that can be given a code. A unit can be as small as a word. Content analysis, a special type of qualitative research, often uses words as the unit of meaning. Most research, however, uses a larger unit such as a meaning (a phrase, sentence, or longer unit that conveys one basic meaning or idea), a sentence, a paragraph, or an exchange (e.g., everything an interviewee said in response to a particular question). Selecting the unit of analysis is important, but it is not a decision that lends itself to a set of standardized rules. The unit you select should make sense given your data and the purpose of the research.

Decisions about codes and the unit of analysis are generally made tentatively. They may be changed as you work with and gain a better understanding of the data. The Ethnograph does not help with the initial development of codes, but it does allow you some leeway in selecting the size of the unit of analysis. Once you have your initial set of codes, the computer will keep track of them for you. Ethnograph has several commands for entering new codes, modifying existing codes, and deleting codes. It also has some annoying limitations. A basic one is the length of code names (no more than ten characters). You can add a paragraph of description to each code, but the limit of ten characters for the name can be irritating when a 12- or 15-character name would be so much more meaningful.

When you have your initial set of codes, you can display them on the screen while you code the file. To code a segment of text in the data file, you can highlight it with your mouse cursor, and then pick the name of the code you want to use. This approach works well when you start your research with at least a tentative set of codes. You can still add new codes and modify old ones if you use this method.

Ethnograph also supports another way to both develop and use codes. You can begin working with a data file and code it as you go. For example, if you decide a passage should be coded *Helplessness*, you can assign that code to the passage. (But you will have to use a shorter name such as *Helpless*.) The next time you find a passage that should be coded *Helpless*, the code is ready to use; you don't have to type it again. Ethnograph adds new codes to its list automatically. This emergent approach to code development appeals to many experienced qualitative researchers, and Ethnograph is flexible enough to support it. Novices may prefer to begin with a small set of broad codes and then gradually make them more precise and complex as data analysis progresses.

There is another annoying limitation of this program that has to do with coding. When you bring a data file up on the screen, the program numbers each line (up to 42 characters). Codes can apply to as small a unit as one line and to as large a unit as the whole data file. You cannot code individual words or phrases; well, you can, but everything else on the line with that word or phrase will be coded the same. Coding by line number takes us back to the old mainframe programs of the 1970s, which were decidedly unfriendly beasts to use. A more modern approach would be to let the users highlight text–a word, phrase, sentence, paragraph, or whatever–then click on a code in a list. Ethnograph's line-oriented approach is only one of several characteristics that remind you of old mainframe computer programs. It has a heavy, "clunky" feel to it, and it is not an "easy to learn" program.

Finding Patterns

Once the data are coded, there are many ways the data can be approached. Ethnograph's data analysis features emphasize the ability to retrieve data associated with codes and the ability to find data that has certain code patterns. Ethnograph supports several types of searches. The simplest, a single code search, involves specifying the code and then reading the text segments the search finds. The program will display all the data units that have the code you specified. Multiple code searches are also supported. You can search for HELPLESS and AGGRESSIVE to find all the segments that were coded both Helpless and Aggressive. Ethnograph also supports more sophisticated searches that include Boolean operators. For example, a search for

HELPLESS and AGGRESSIVE not FEMALE

would find all the segments coded helpless and aggressive that were not coded female.

The thought may have occurred to you that if the multiple code searches are limited to finding two or more codes in the same segment or unit of data, this is not very useful because many types of analysis will involve looking for two or more codes that may be in close proximity to each other but not in the same data unit. Fortunately, Ethnograph has a proximity option. You can tell the program to look for codes that appear within a certain number of lines of each other. You can also specify the sequence of occurrence if you are only interested, for example, if CODEA appears before CODEB.

Another problem qualitative researchers face with manual methods is that the coded data are taken out of context. It is hard to put text coded a certain way back into the context so you can easily see what was being said before and after that segment. Ethnograph has options that let you see a little or a lot of the surrounding text. This is useful, but it may also mean you generate many pages of output when you print the results of a particular search.

Developing Theories/Understanding. Some researchers distinguish between looking for patterns and developing theories or understanding based on the data. Ethnograph supports searches for basic and sophisticated patterns but does not have additional features designed specifically for theory building or the development of understanding (verstehen).

Writing Your Report

The final step in a research project, write-up, is not supported by Ethnograph.

Ethnograph's Advanced Features

There are several things Ethnograph does that takes code-and-retrieve work beyond what can be done conveniently with manual methods. For example, it supports nested or tree coding categories. You can create a tree of codes, with each branch of the tree representing a major type of code. One branch might be Aggression. Off that branch might be more detailed codes such as Physical, Verbal, and Passive. Up to seven levels of coding are supported. Ethnograph can display the branches in your tree of codes in text format (graphic would be better). The program will also let you search for a "branch" of the tree like Aggression and include all the code categories on that branch (such as physical, verbal, and passive).

Ethnograph also lets you code the same segment with as many as 12 different codes, a feature that is hard to duplicate with manual coding. You could then, for example, look at all the segments coded for Aggression and Adolescence and Physical.

Another nice feature of Ethnograph is the ability to conduct searches that include identifiers such as the name of the person who said something or the case number. You could, for example, find every utterance of BILL that was coded AGGRESSIVE. Files can also have codes, which means that you could do a search in all the files of interviews from patients with depression, or children with a specific type of learning disability. This ability to search other than the actual data applies as well to memos you may write and attach to data. The memos are kept separately and can be printed out by themselves, but they can also be included in a search.

Finally, Ethnograph can provide summary data on the frequency of each of the codes you use. This type of data can be generated for individual data files or for sets of files.

In summary, Ethnograph is a basic, but powerful tool for researchers who want to use a code-and-retrieve approach with their data. It does the job well, and has a number of advanced features. On the negative side, it is a bit awkward to use. The user interface would have been outstanding compared to mainframe software in the early 1980s, but when compared to programs today, it seems ungainly, clumsy, and confusing.

We began our review of qualitative software with Ethnograph because it was one of the first packages available, and it illustrates what should, at a minimum, be available in a program designed to support code-and-retrieve data analysis. The next program is an improvement on Ethnograph in two ways–it is more powerful and it is easier to use.

NUD*IST 4.0: AN EXAMPLE OF CURRENT QUALITATIVE SOFTWARE

The developers probably picked the name of this program–NUD*IST–because it would attract attention and be remembered. The name stands for *Non numerical Unstructured Data by simple but powerful processes of Indexing Searching and Theory-building*, but the authors probably came up with the name first and then found a phrase that fit.

NUD*IST is probably the most widely used qualitative data analysis program today. There are several good reasons. It is a relatively powerful code-and-retrieve program. In virtually every area it outshines Ethnograph. You can assign an unlimited number of codes to a data unit, for example, while Ethnograph accepts no more than 12. When searching, NUD*IST has more than twice the operators available to detail exactly what sort of search you want to run.

Also, while it is not easy to learn to use, it does come with a number of resources for mastering the program, including a manual, an online help system, and an online tutorial that will walk you through all the steps in a focus group/interview study of office staff in a small company that is making changes in the smoking policy. In reviewing this program, we will divide the work into the same seven categories that were introduced in the review of Ethnograph.

Conception and Planning of the Study

As with Ethnograph, NUD*IST offers no help here.

Data Collection

NUD*IST does not have any data collection features. It enters the picture when you have data to be entered into files the program can use. Like Ethnograph, NUD*IST has some requirements for text files it will accept. Data files should generally have a line at the beginning that starts with an asterisk (*). NUD*IST will treat it as a header that contains information about the file. You can also divide data files into sections by placing an asterisk (*) at the beginning of a line and then typing a subheading that describes the section. Files ready to be imported into NUD*IST should be "plain text" or ASCII files.

Conversion of Data Files

NUD*IST is divided into two subsystems–a document system and an index (coding) system. The document system keeps track of all the documents in your study. These documents can be online (e.g., a transcript of an interview that has been imported into a NUD*IST file), external (the document system can refer to an external data source such as a movie or a set of photographs), or memos (material you create that is connected or related to your data).

While there are ways to use NUD*IST to code multimedia documents such as movies, we will focus on the way text data in a NUD*IST data file is handled. The program can input data from word-processing programs, but only as plain text. You cannot import formatting such as different font sizes, bold and italics, and so on.

NUD*IST uses windows for different types of work. The Document Explorer window is where you import documents, as well as edit, revise, and annotate them. NUD*IST does have a built-in editor, but it will be used primarily to edit files already imported and to create memos you attach to data files. The most common way to bring a file into NUD*IST is to create it in a word processor, save it as a "text" file, and then use NUD*IST's Import command. Files you import are listed in the Document Explorer window. The header you put at the beginning of the file is also displayed when you click the file name. Other commands in the Document Window let you add memos to a data file, browse files, check the properties of files (e.g., whether it has been coded, etc.), and generate a report that includes the document.

If you use NUD*IST regularly and import many different files, the program has a macro feature that automates the file import process. NUD*IST automatically creates a folder on your disk drive named RAWFILES when you begin a new project. It is relatively easy to copy all the files you want to import into the RAWFILES folder and then run a macro that adds all of them into NUD*IST file structure.

Coding the Data

Once you have your data files in NUD*IST, you can begin the next stage, coding (NUD*IST sometimes uses the term *indexing* instead of coding). Coding involves two steps, setting up the categories and then assigning data segments to appropriate codes. NUD*IST calls the code categories "nodes" but it refers to the process of assigning data units to categories as coding.

Each node (code category) has a title or name. You can also add extended definitions of nodes (codes). NUD*IST uses nodes in the same way other programs use code categories, but they can also be used for more general purposes. You could create a note, for example, of all the interviews from students in the sixth grade, while another node might be for the interviews from the sixth grade teachers. Nodes can also represent things like who the interviewer was, or the date the data were collected. As you might expect, any unit of data can be coded many different ways.

NUD*IST automatically creates some nodes (code categories) including one named *Document Annotations* where notes you add to the data file are stored. To say that nodes "contain" this or that is not quite correct. Nodes are actually sets of pointers that identify the location of selected parts of the data.

When you create nodes, you tell NUD*IST whether you want it to be a "free node" or part of the Index Tree. The Index Tree organizes the nodes into parent/child relationships. For example, you could have a node named Sixth Grade, and under that, a Parent node with nodes for Teacher interviews, Parent interviews, and Student interviews.

One advantage of NUD*IST for researchers taking an emergent approach to data analysis is the flexibility of the tree structure. You can move nodes around in virtually any way you want and make changes in the parent/child structure as you go.

Once you have created your nodes (code categories) you can begin to add codes to the data. (Actually, you can do it in reverse–read the data and code the data as you go. NUD*IST will automatically create the nodes–code categories–as you use them.) To add codes to text, you use the Browse command to display a file. NUD*IST also displays a palette of commands for adding and deleting codes, among other things. The entire process is menu driven. For example, when you highlight a unit of data and click the Add Coding command, a dialog box pops up that allows you to type the name of a code category or pick a code from a list.

Highlighting is the principle means of selecting a segment of text for coding. However, what you highlight is not the unit that is coded. The smallest unit of data is the paragraph. Thus, if you highlight the words "was afraid" in a paragraph with 50 words, the coding will be applied to the entire paragraph, not just the two words you highlighted. You can, of course, break those two words out into their own separate paragraph if you want to code only those two words, but the result will be a data file that is difficult to read.

Finding Patterns

Like Ethnograph, NUD*IST will find all the data units that have a certain code. This is called an Index Search (as compared to a text search which is described below). And, again like Ethnograph, the program can do searches based on several codes as well as searches in only one part of the data (e.g., teacher interviews but not student interviews).

NUD*IST has options for precisely defining the search you want to perform. The options include typical Boolean operators such as AND, OR, and NOT. There are many others, and the result is a powerful way of looking for complex patterns in the data. There are, for example, ways of searching for code patterns in data units that are close to each other, the codes need not be in the same data unit.

With NUD*IST you can also search any file, set of files, or node (code category) for a word or phrase. The Search Text command in NUD*IST is more powerful than the typical search functions in word processors, and you can limit the search to certain files or nodes. The program shows you the *hits* of text searches, but it does more. It creates a new node based on your search. For example, if you search for the words "my teacher" in interviews with students, NUD*IST creates a *My Teacher* node. You can then use that node like any other. It can, for example, be incorporated into the tree structure.

The Index Tree in NUD*IST is another way of looking for and highlighting patterns. You can organize code categories into tree structures so that a global concept (such as Attitudes Towards Smoking) are broken down into subcategories (such as Positive and Negative). NUD*IST has more flexibility than Ethnograph when it comes to moving code categories around in the tree structure, and the visual displays of the relationships between the code categories are more informative. You can even rearrange the tree structure of the codes by dragging nodes from one location to another in the visual displays. The ability to visualize relationships between categories and to change those relationships is a powerful feature of NUD*IST.

Another advanced feature of NUD*IST is the ability to look at data associated with a particular code in context. You can, for example, look at the four paragraphs before and after a data unit so that you can see the context in which the coded data occurred. NUD*IST also has a Jump to Source command that will take you from the data segment to the document the segment came from. You can then browse the entire document if you wish.

Developing Theories/Understanding

NUD*IST is, in essence, a code-and-retrieve program, but some of the advanced features support developing theoretical explanations of the phenomena you are studying. For example, the Index Search command is very powerful, and it has options that let you explore tentative hypotheses in some general ways and then narrow or focus your search to confirm your guess. For example, if you end up with a series of questions (searches) that address a particular hypothesis, they can be arranged in the Index Tree according to the relationships between the questions.

Another way NUD*IST supports theory development is the Index Tree. Because you can organize and reorganize nodes, it is relatively easy to create subordinate and superordinate representations of the relationships between different nodes and then explore the implications of those hypothesized relationships using NUD*IST.

NUD*IST has stronger support for developing theory than for developing understanding, but there is some support for more holistic efforts to analyze the data. The ability to look at data units in context and to conduct searches that take into consideration surrounding data are two examples. Another is the Collect command that lets you put together data from two or more nodes to create another node.

Advanced Features

Several of the sophisticated features of NUD*IST have already been mentioned. Some that have not involve the quantification of data. NUD*IST can provide several types of summary data such as the number of times codes occur in various settings (e.g., the number of positive opinions about issue X expressed by male versus female interviewees). The program can also output tables that summarize the data, and it can export data files to statistical analysis packages such as SPSS.

Writing Your Report

Like Ethnograph, NUD*IST provides no support for write-up except for the ability to create summary tables that can be included in reports.

In summary, NUD*IST is a powerful and relatively easy to use program that supports sophisticated code-and-retrieve data analysis. Unlike Ethnograph, which has the feel of a 1970s mainframe program, NUD*IST makes good use of windows to organize work and it takes advantage of menus and lists of commands to help you make decisions about what to do next. It is not as intuitive and straight-forward as the current crop of sophisticated word processors. However, word processors have about twenty years of maturity on programs like NUD*IST. This program will likely become even more powerful and easy to use as new versions become available.

The program can be used at a simple level that does not go beyond the basics of code and retrieve, but it has many advanced features that can enhance this type of data analysis. Unfortunately, those advanced features are not always obvious to the novice user, and their use is not always intuitive. NUD*IST comes with a four-lesson tutorial that does cover many of the advanced features. If you are new to NUD*IST, completing this tutorial is highly recommended. Some of the advanced features, such as the creation of macros to code and organize data, require considerable effort to learn. Once learned, NUD*IST's advanced features will be very helpful for certain types of data analysis.

ATLAS/TI VERSION 4.1: A NEW FORM OF QUALITATIVE DATA ANALYSIS SOFTWARE

The original version of Atlas/ti was developed at the Technical University of Berlin in 1989. Thomas Muhr (1997), the social scientist who created Atlas/ti, describes it as a "powerful workbench for the qualitative analysis of large bodies of textual, graphical, and audio data" (online). That is probably a good way of thinking about this program–as a workbench with many different tools for different types of work.

The workbench metaphor, however, is less central to the way this program works than the acronym VISE. VISE is short for Visualization, Integration, Serendipity, and Exploration. Muhr believes qualitative data analysis should make heavy use of visualization because that is the way humans think. Atlas/ti has many tools for creating and modifying visual representations of questions, problems, and relationships. The program itself is also object oriented–it uses windows, clickable options, and graphics instead of command lines. The second principle, integration, emphasizes that even when you break down data you should always keep the whole in mind and relate the parts to that whole. Muhr uses some novel terms in Atlas/ti to maintain that emphasis on integration. For example, a core concept of the program is the Hermeneutic Unit. The meaning of Hermeneutic Unit is similar to the concept of Project in Ethnograph. A Hermeneutic Unit is simply all the files associated with one research project.

The third principle, serendipity, emphasizes the unplanned nature of many of our discoveries in qualitative research. We do not propose precise hypotheses before the research begins and then gather and analyze data specifically to support or refute that hypothesis. Our hypotheses, guesses, and hunches may emerge from our engagement with the data. Atlas/ti has a number of tools for browsing or exploring the data, including powerful hypertext functions that go well beyond what is available in either Ethnograph or NUD*IST.

Exploration is the final principle, which Muhr selected because he wanted the principles to spell a word. After VIS, he needed an E to make VISE. Exploration is really a summary of the other three. VISE thus represents a methodological framework that is a bit more holistic than the underlying frameworks for either Ethnograph or NUD*IST. It encourages a less structured, more open, discovery approach to the data.

Atlas/ti's framework is actually more distinct from that of Ethnograph and NUD*IST than the reality of the program. It is still, at the core, a code-and-retrieve program. However, around that core are some powerful tools that can move you well beyond basic code-and-retrieve data analysis.

Conception and Planning of the Study

As with the other two programs reviewed, Atlas/ti does not support this phase of research.

Data Collection

There are no special features to support data collection.

Conversion of Data Files

Atlas/ti has more built-in features for handling data such as photographs, x-rays, movies, audio, and video than the other two programs. You can actually import audio files, for example, and code them as you would a text file. Then, when you want to "read" all the data segments that were assigned a certain code, you will hear the audio instead of see the text.

However, in this review, we will assume the data are in relatively standard text files. You begin work with Atlas/ti by specifying the name of a new hermeneutic unit, which is just a fancy name for project. You don't copy your files into a single directory, however. Instead, the hermeneutic unit points to the location of all the relevant data files–the primary documents. The files can be on your hard drive or somewhere on a network you're connected to.

Coding the Data

Once you have identified all the data files, the next step is coding. Atlas/ti supports four different types of coding: open coding, in-vivo coding, code-by-list, and quick coding. Quick coding is a way of specifying a code and then going through a document and selecting one segment after another you want to assign that code to.

You may be wondering what the size of the data segment is in Atlas/ti. Unlike both Ethnograph and NUD*IST, the unit size is variable. It can be as short or as long as you like. If you assign a code to four words in a ten-word sentence, those are the only words in that sentence that carry the code. A data unit can be as small as one letter and as long as an entire document.

The other three methods of coding all require you to display the data file in a window, highlight a segment of the data with your mouse cursor, and then select or create a code. It sounds simple and it is. If you "code by list" a list of codes pops up and you just click the one you want to use. Atlas/ti doesn't require it, but if you have an idea about the codes you want to use, it will accept "free codes" anytime. They are codes you haven't yet assigned to any segment of data. Free codes are added to the list Atlas/ti keeps of acceptable codes.

If you do "open coding" (by clicking an icon on the left side of the screen) a text box pops up and you type the name of a new code after highlighting the data you want to code. The last type of coding, in-vivo, is an odd command that lets you pick text to be coded and then use the text as the name of the code. Thus, the same material serves as the name of the code category and as the content of the category as well.

While Atlas/ti is not easy to get the hang of, once you do it is one of the easiest programs to use for coding. You can also easily write annotations and memos. As you might expect, there are commands for renaming and deleting codes, and for removing a code from a data segment.

Finding Patterns

As you code your data, Atlas/ti keeps all sorts of data on the coding patterns. Drop-down menus at the top of the main window list all the segments that have been coded, and the codes assigned to the segments. Another drop-down menu lists all the codes you've created thus far for this project (hermeneutic unit). However, the easiest to understand and simplest to use tool is the "margin area" of the screen. The margin area on the right side of the window has a graphical indication of each code in the document along with a one-word description. Colored, squiggly vertical lines tell you where that data segment begins and ends. This simple but brilliant graphic display tells you a great deal about the way your data has been coded to date. It is far superior to anything in either Ethnograph or NUD*IST.

This is only a sample of the types of support for pattern analysis that Atlas/ti has. It all goes to help you read your data and your coding and to think through your data analysis. Atlas/ti uses the term *coding* to talk about starting with segments of data and assigning categories (codes). When you want to go in the opposite direction–from codes to data–that is termed *decoding*. Simple retrieval of all the text with a particular code is easy. Pull down the menu of codes and click one. Two things will happen. First, in the main data window the first instance of that code will be displayed in white-on-black text. The surrounding text is displayed normally. Also, a list of all the instances of that code in the data file will pop up. You can click any data segment in that list and the main data window shows that part of your data. Again, this is simpler and more intuitive in Atlas/ti than in the other two programs. Most people will probably use this click and search method to do simple retrievals, but like the other programs you can print reports that output all the instances of a code to a file or printer.

Like NUD*IST, Atlas/ti has a text search tool that is much more powerful than the standard search tools in word processors. Even more appealing, however, is QueryTool.

Complex searches require using the QueryTool, a powerful way of conducting searches that can involve several types of relationships between a number of codes. In fact, the QueryTool may be overkill for most of the simple types of code-and-retrieve work reported in the literature today. It can create very complex search strategies that would be virtually impossible to do manually. QueryTool can locate coded data based on three types of relationships:

Boolean: AND, OR, XOR, and NOT.

Semantic: As with NUD*IST you can create meaningful relationships between codes. There can, for example, be parent codes and sibling codes. Atlas/ti has commands (actually "operators") for doing searches that look for SUB codes (codes that are below a particular code in a hierarchy), UP codes (all codes above another code in a network), and SIBLING codes (all codes that are children, or grandchildren and so on, of the same parent as another code).

Proximity: You can search for sets of codes that have different spatial relationships. For example, you can specify how far apart the codes can be, whether one is embedded in a data unit for another code, whether the data for two codes overlap, and whether the same data unit has two codes (co-occurrence).

Atlas/ti can look for these types of relationships among two or more codes. It can also look for them among supercodes and code families. A *supercode* is simply a query you have already done. If you supply a name for this search in the supercode text box, you have also created a supercode. The query logic of that search becomes the beginning point for your next search when you start with the supercode (name) of that search. A *code family*, which can have a separate name, is a set of codes connected by a boolean OR operator. Thus the A *code family* might be made up of code a1 OR a2 OR a3.

The search features of Atlas/ti are as impressive as those of NUD*IST. They are, however, probably harder to learn and more clumsy to use. This is by far the most difficult of the basic features to use in this program. Queries, like other work, have all sorts of support including clickable command icons. But QueryTool is still not a very friendly tool. One part of the QueryTool, for example, works like the stack in a microprocessor. (A stack is a section of memory inside the central processing chip where numbers that will be used soon are temporarily stored). In Atlas/ti there are commands like clear, swap, and push, which are common commands for manipulating the stack in a computer chip. If you happen to program microprocessors in machine or assembly language, this will be a snap for you. If you are one of the 99.999% of qualitative researchers who don't have that as a hobby, it will not be as intuitive and you may spend some time learning how to do it.

Developing Theories/Understanding

Tools like supercodes were designed to support on-the-fly development of theories. They can be conduits for testing progressively more precise guesses about relationships among codes in your data. However, the main theory building tools in Atlas/ti are in an area called Networking. Networking tools help you build visual models of relationships in the data and then test the data to see if those relationships exist. Basically, Atlas/ti supports the creation of networks that include nodes and links. A network you create can have several types of nodes and links. There are also several ways of displaying these networks but the emphasis is on visual rather than textual modes.

Atlas/ti uses special mathematical symbols to represent different types of relationships. For example, the statement "CodeA == CodeB" means *Code A is associated with Code B*. "CodeA =>, C CodeB" means *Code A is the cause of Code B*. "CodeA <> CodeB" *means Code A contradicts Code B*. Familiarity with the field in discrete mathematics called *graphing theory* helps when you're working with network building in Atlas/ti. This is an advanced feature of the program, but it has a steep learning curve and the tutorial and learning materials, while helpful, do not make mastering this part of the program enjoyable. The title of one section of a quick-start lesson that comes with Atlas/ti, "Don't be afraid of the QueryTool," suggests the authors are aware of this but expect users to put in the effort to learn it. It is unfortunate that the developers based the QueryTool on metaphors unfamiliar to most qualitative researchers. It could have been based on any of several other much more comfortable and familiar metaphors.

Many, but not all, of the features that support theory development in Atlas/ti are available in NUD*IST in one form or another. The most important area where that is not true is in the graphical representation of networks of relationships. Atlas/ti is far superior to NUD*IST in this area. However, that superiority comes with a cost. Atlas/ti is not an easy program to learn. Expect to spend many hours with this program before you can take advantage of the advanced graphics features.

All the features mentioned thus far support a grounded theory approach to building general or local theory. They can be used for more holistic or hermeneutic approaches, but they were not specifically designed for that purpose. Another tool in Atlas/ti is a better fit. Holistic approaches tend to look for links between different parts of the data rather than code-based relationships. Atlas/ti's hypertext tools will support a holistic approach to data analysis. You can create hypertext links in your data that take you to all sorts of supporting data, including:

1. Points where similar opinions or results are reported
2. Support for statements or conclusions
3. Contradictory material
4. Alternative explanations of similar events

Atlas/ti supports a sophisticated form of hypertext linking that lets you specify links as well as what type of link it is. That is, the type of relationship in the hypertext link can be *named*. That will allow you to follow all the support links at one point and later follow all the contradictory links for a particular issue or hypothesis. Atlas/ti can also display the hypertext links graphically, something even powerful Internet browsers can't do (though some of the programs for creating Web pages do). The graphical presentations are also active–you can click a hypertext node in a graphic and go to it.

Atlas/ti supports several different ways to create hypertext links. Some overlap with the ways of creating and searching codes and some are specifically for hypertext. This feature of Atlas/ti is the closest any of the three programs come to strong support for a holistic approach to qualitative data analysis.

Writing Your Report

Atlas/ti does have a built-in editor, which is good enough for writing small notes, memos, and annotations, but you would not want to use it to write your final report. However, the program is capable of producing graphical representations of concepts, relationships, and theoretical frameworks that can be output and used in final publications.

A much more significant capability of Atlas/ti in terms of publishing your work is the ability to output a hermeneutical unit (all the files associated with a research project) as a World Wide Web site. You could, for example, put your entire study, including data, codes, and hypertext links, on your own Web site. The output file will include the graphics as well as the text in your files. The output from Atlas/ti is not always perfect; it may require some hand tweaking before it works properly on your Web page, but it makes putting your project online something to consider versus something that is simply too time consuming to even think about.

Advanced Features

Many of the advanced features of Atlas/ti have already been mentioned, but there are several others. An important one for researchers who will be working in teams is the ability to keep track of more than one researcher who works on the data. If you are working with a team of two or more researchers, each can go through the data, add codes, make annotations, and write memos. The work of each researcher is organized and stored separately. If you want to look only at what Researcher 1 is doing you can; or, if you want to be able to distinguish between the coding of Researcher 1 and Researcher 2, you can.

Finally, like NUD*IST Atlas/ti can output several types of data files to statistical analysis programs like SPSS.

To summarize, Atlas/ti is the most powerful and most sophisticated of the three packages reviewed. In virtually every aspect of the program, it makes heavy use of a graphical user interface, menus, windows, and icons to make using the program as easy as possible. All that effort has helped. Basic operations are easy. Unfortunately, the parts of the program that are the most unique and most useful for advanced qualitative research are still complex, hard to learn, and hard to use. This program will probably appeal to people who recognize the need for advanced, graphically-rich data analysis work and who are willing to invest the considerable time and effort required to become a proficient user of Atlas/ti. Atlas/ti is a glimpse into the future. It hints at what is coming, but the program itself has some serious flaws.

WHICH PROGRAM FITS YOU AND YOUR NEEDS?

The three programs reviewed thus far are by far the most widely used today. Of the three, we believe Ethnograph is the weakest and should probably be dropped from consideration unless you already know the program well and it supports the type of work you want to do.

Between NUD*IST and Atlas/ti, the choice is more difficult. It is not just a matter of choosing which one has the most support for the type of data analysis you want to do. If that were the only criterion, Atlas/ti would probably win most comparisons. However, the learning curve for Atlas/ti and its complexity of use must be considered. If you are doing a one-shot study, are working under a tight time line, and will not likely do this type of research regularly, NUD*IST would be a better option if it supported the type of data analysis to be done. NUD*IST is not easy to learn either, but it can be mastered in far less time than Atlas/ti. Where both programs will help with the data analysis you want to do, Atlas/ti would probably be the preferred program

1. when you have plenty of time (and energy) to learn to use it before you feel the pressure to start data analysis, and
2. you will be doing this type of analysis regularly, and
3. the advanced features of Atlas/ti will add significantly to your work.
4. or where it is absolutely necessary that you have a flexible data unit that is as short as a word since NUD*IST cannot code anything shorter than a paragraph.

Otherwise, NUD*IST may be a better choice.

Another situation that favors Atlas/ti is when you want to do holistic analysis and the hypertext features of the program are appealing. You need not learn all the complexities of the QueryTool to use the hypertext features, and Atlas/ti's ability to output files that can be put on the World Wide Web is another plus.

Ethnograph, NUD*IST, and Atlas/ti are the three most popular qualitative data analysis programs today, but there are many others that work much like these three. The list of resources at the end of this paper details sources of information on about 80 other programs currently available. Before ending this discussion of software, however, we would like to mention one additional type of software.

SPECIALIZED QUALITATIVE DATA ANALYSIS SOFTWARE

All three of the programs reviewed in this paper are general purpose programs that support a code-and-retrieve approach to qualitative research. Ethnograph, NUD*IST, and Atlas/ti are by far the best known and most widely used of the 70 or so programs currently available to qualitative researchers. Quite a few of the other programs in use today are commercial packages that were created for other applications. Word processors, for example, like Microsoft Word and WordPerfect, are often used for basic qualitative data analysis. Database programs like FOLIO, are also used by a number of scholars.

General purpose qualitative software packages, and re-purposed commercial software are common in qualitative data analysis, but there is another type of software that deserves attention–specialized or special purpose programs. There are many types of specialized qualitative research–such as content analysis and semiotic research–and for many of those specialized types of research there are one or more computer programs that were created specifically to support that type of work. Code-A-Text *(http://www.codeatext.u-net.com/)* for example, is a program developed to analyze discourse. Several programs are available for a popular form of qualitative data analysis called content analysis. Content analysis looks essentially at the occurrence of words and phrases in text. Some of the programs for content analysis use pre-existing coding schemes to assign common words to categories (e.g., mad, angry, hostile, hurt, and so on might be coded "emotion"). For a discussion of software for content analysis see Bell (1997) and the "Software for Content Analysis" *(http://www.gsu.edu/~wwwcom/content/csoftware/software_menu.html)* section of the Content Analysis Resources Web page at George Washington University *(http://www.gsu.edu/~wwwcom/content.html)*.

The programs mentioned here are only a sample of those currently available (and many new ones appear each year). The best way to get information on software for special types of qualitative data analysis is to do a search on the Internet (e.g., search for content analysis and software) or to go to some of the Web sites listed at the end of this paper that keep track of software for qualitative data analysis.

HOW NARROW IS CODE-AND-RETRIEVE SOFTWARE?

The three programs reviewed in detail in this paper all take a somewhat atomistic approach to data analysis. (Atlas/ti does provide some support for holistic analysis, but even this program has more features and tools for atomistic analysis.) Computer programs like Ethnograph have made atomistic analysis easier and they have made it more sophisticated. There are things you can do with all three of the general purpose programs reviewed earlier that would be difficult or impossible with the old cut-and-paste method.

Naturally, if you are worried about the influence of code-and-retrieve approaches, a software program that helps you do even more coding and retrieving will not be especially welcome. Some (Barry, 1998; Mason, 1996) feel the power of code-and-retrieve software may mislead researchers into thinking the codes and therefore relationships between codes represent "real" (concrete) things instead of constructions of the researcher. Other critics have argued that we are automating something that should not be done at all in many studies because it supports one approach–grounded theory, which is not always appropriate (Coffey, Holbrook, & Atkinson, 1996). In essence, Coffey and her colleagues argue that programs like NUD*IST were created specifically to support an approach that involves:

1. Coding the data
2. Developing a tentative theory
3. Testing the theory by looking at relationships between codes
4. Revising the theory based on the results
5. Testing the revised theory by looking at additional code relationships

This is the essence of grounded-theory research, and some feel this is not always the appropriate framework for qualitative research. It is clearly a way of looking for general or local theory as opposed to storytelling, but if you are more concerned with understanding or with storytelling, is a program like NUD*IST appropriate? If the only way the software can be used is to support grounded-theory research, then people doing narrative research or verstehen research should not use it. Not everyone agrees with the assumption that these programs only support grounded theory research, however.

Kelle (1997) has articulated another perspective on code and retrieve data analysis. For him "code and retrieve represents an 'open technology' applicable in various theoretical and methodological contexts" (Online). Kelle outlines two approaches to using code-and-retrieve data analysis, one that fits the atomistic approach already discussed and one that fits a more holistic approach. Table 1 summarizes the major points he makes.

TABLE 1
Two Approaches to Using Code-and-Retrieve Data Analysis Methods

	Holistic/Interpretive	Atomistic/Empirical
Codes are	efforts to make meaning tentative, emergent, theory-laden	facts that lead to theories logically, objectively derived
Codes are created	as you go along	before data analysis
Codes can be used to	build understanding Test guesses, hunches	test hypotheses derived from existing/emerging theories
Hypotheses (guesses) are	tentative, imprecise Conjectures	empirically testable statements

We agree with Kelle that code-and-retrieve methods of data analysis do not lead inevitably to the adoption of a grounded-theory approach, or the creation of general theory and the treatment of the data as factual. In Kelle's (1997) words,

> The coding of text does not serve to condense relevant information and to decide whether a certain person or event falls under a certain class of events or persons, but simply to make sure 'that all relevant data can be brought to bear on a point.' Here, the function of coding is restricted to sign-posting: codes are stored together with the 'address' of a certain text passage and, drawing on the information, the researcher can locate all the possible information provided by the textual data on the relevant topic. . . . such qualitative codes which represent 'perspectives' and which serve as 'sign-posts' are not very useful to test empirical statements about the exact relationship between two defined variables. (Online)

We recommend Kelle's paper to anyone who is concerned about the place of coding and retrieval in qualitative data analysis. It is clearly an important part of atomistic approaches, which tend to emphasize the development and testing of theories in much the same manner as traditional quantitative research. However, many holistic researchers also use code-and-retrieve, but in a different way. Holistic researchers tend to use coding to help them organize their thoughts about the data.

SUMMARY

Most of the existing group of general purpose qualitative data analysis programs support a code-and-retrieve approach to data analysis. Researchers who do research for purposes such as to develop general or local theory will find many of these programs appealing.

Some scholars believe this emphasis on one approach to data analysis is part of a pattern–computers in qualitative research support grounded-theory research over other methods such as narrative and interpretive approaches. They are concerned that computer programs make it so easy to do grounded-theory work that novices may be seduced into using that framework even when it is inappropriate. For example, if the purpose of the research is interpretive or narrative (storytelling) most of the current crop of programs have to be adapted to support this type of work.

Another group of scholars agrees that there is a danger that computer programs make it easy to use inappropriate methods, but they argue that computer software can be used to support many types of data analysis. We agree somewhat with both groups. Programs like Ethnograph, NUD*IST, and Atlas/ti do have built-in support for grounded-theory methods, and researchers should think through their ideas about methodology before selecting a computer program to help them with data analysis. However, none of the programs are really that appealing–they are not easy to learn, and the powerful features of most require considerable effort to master. We think the threat of computers as seductive influences that will lead us astray is overrated. On the other hand, the general milieu of qualitative computer-based data analysis emphasizes the atomistic at the expense of the holistic. Researchers who wish to take a holistic approach to analysis will have to work hard at keeping that in mind because they will have to adapt and adjust software to support holistic work (and look for features in the program that are not always highlighted or illustrated in the documentation). For holistic researchers, programs like Atlas/ti with enhanced hypertext features are probably the most promising tools.

Finally, while we reject the demonization of computers in qualitative research, those who raise serious concerns do have an important point. Computers make it very easy to break data down according to codes created by the researcher. Once they are enthroned in the corpus of the computer, it is very easy to think of them as real, concrete things rather than what they are–socially constructed, subjective perspectives. Anyone who uses a computer to facilitate qualitative data analysis should keep that in mind.

REFERENCES

Bakeman, R., & Gottman, J. M. (1986). *Observing interaction: An introduction to sequential analysis*. Cambridge: Cambridge University Press.

Barry, C. (1998). Choosing qualitative data analysis software: Atlas/ti and Nudist compared. *Sociological Research Online 3*(3) [On-line]. Available: *http://www.socresonline.org.uk/socresonline/3/3/4.html*

Bell, B. (1997). Qualitative analysis: Web & software resources. [On-line]. Available: *http://ihs2.unn.ac.uk:8080/bbquall.htm*

Bodgan, R., & Biklen, S. (1992). *Qualitative research in education* (2nd ed.). Boston: Allyn & Bacon.

Coffey, A., Holbrook, B., & Atkinson, P. (1996). Qualitative data analysis: Technologies and representations. *Sociological Research Online, 1*(1) [On-line]. Available: *http://www.socresonline.org.uk/socresonline/1/1/4.html*

Collins, P. (1998). Negotiating selves: Reflections on 'unstructured' interviewing. *Sociological Research Online, 3*(3) [On-line]. Available: *http://www.socresonline.org.uk/socresonline/3/3/2.html*

Feyerabend, P. (1981). *Philosophical Papers*. New York: Cambridge University Press.

Kelle, U. (1997). Theory building in qualitative research and computer programs for the management of textual data. *Sociological Research Online, 2*(2) [On-line]. Available: *http://www.socresonline.org.uk/socresonline/2/2/1.html*

Kuczynski, L. (1998). Reflections of a closet qualitative researcher. *The genetic epistemologist: The journal of the Jean Piaget society, 26*(2) [On-line]. Available: *http://www.piaget.org/GE/1998/GE-26-2.html*

Lonkila, M. (1995). Grounded theory as an emerging paradigm for computer-assisted qualitative data analysis. In U. Kelle (Ed.), (p. 41-51), *Computer-aided qualitative data analysis: Theory, methods, and practice*. London: Sage.

Miles, M., & Huberman, M. (1994a). *Qualitative data analysis* Newbury Park, CA: Sage.

Miles, M. and Huberman, M. (1994b). Data management and analysis methods. In N.K. Denzin, & Y.S. Lincoln (Eds.). *Handbook of qualitative research* (pp. 428-444). Thousand Oaks, CA: Sage.

Mason, J. (1996). *Qualitative researching*. Thousand Oaks: Sage.

Muhr, T. (1997). *Atlas/ti user's manual and reference*. Thousand Oaks: Sage/Scolari.

Mulder, J. (1994). The mechanics of qualitative analysis. *Issues in Educational Research, 4*(2) 103-108. [On-line]. Available: *http://cleo.murdoch,edu.au/gen/iier/iier4/942p103.htm*

Smith, J. (1993). *After the demise of empiricism: The problem of judging social and educational inquiry.* New York: Ablex.

Strauss, A., & Corbin, J. (Eds). (1997). *Grounded theory in practice: A collection of readings.* Thousand Oaks: Sage.

Sources of Software Discussed in This Paper

The programs described in this paper were developed all over the world–from Germany (Atlas/ti) to Australia (NUD*IST). In North America most, but not all, programs mentioned in this paper are distributed by Scolari, a division of Sage Publications. Scolari's Web site address is *http://www.scolari.com/* That site has information on many different programs as well as options for downloading demonstration versions of the software. Scolari will also mail a catalog to you on request (phone: 805 499-1325), and will supply demo versions of most of the software they distribute for a small fee.

There are also a number of specialized sites that support users of a particular qualitative software package. A few are listed below:

Ethnograph	*http://www.QualisResearch.com/*
Atlas/ti	*http://www.atlasti.de/*
NUD*IST	*http://qsr.latrobe.edu.au/*

Finally, several individuals and organizations support Web sites that keep track of qualitative software. These sites include original articles and reviews of programs, links to other sites, support for discussion groups about qualitative research and computers, and software archives where you can download demonstration versions of commercial software, and full versions of freeware and shareware programs. Some of the best of this type of site are described here.

QualPage. This general purpose site which is maintained by Judy Norris listed over 20 different computer programs for qualitative analysis in 1999. More will probably have been added by the time you read this. This is one of the best sources for information on qualitative software. The Web page address for the software page is *http://www.ualberta.ca/~jrnorris/qda.html*

CAQDAS Network. The Computer-Assisted Qualitative Data Analysis Software Network was created to disseminate information and expertise about the use of computers in qualitative research. The Web site has many different types of resources including a software download area, a list of relevant publications, a discussion list, help files, descriptions of upcoming events such as conferences, a newsletter archive, and links to many other Web sites. This is another outstanding resource, perhaps the best that concentrates on software for qualitative research. The Web address is *http://www.soc.surrey.ac.uk/caqdas/*

The Qualitative Research Page. This comprehensive site on qualitative research has a section of computer software that is very helpful. The Web address is *http://www.oit.pdx.edu/~kerlinb/qualresearch/*

W. Michael Reed

Assessing the Impact of Computer-Based Learning: A Research Course for Graduate Instructional Technology Programs

The purpose of this article is to explain a research course specially designed for graduate students in instructional technology programs. I developed this course over 10 years ago in response to my frustration of working with graduate students approaching their thesis or dissertation proposal stage and their having difficulty understanding which statistical procedures they should use in their research. Initially I was very critical of the standard research courses for not better preparing my students. As I thought about the problem more, I decided that it was more a result of graduate students in a variety of majors or areas of specialization attending the courses. I decided it was probably difficult for the research professors to tailor the instruction for the varied students, so they understandably presented the statistical information in very general terms, or usually no context especially meaningful to them. To remedy this problem, I decided to design a course that would address the most-often used statistical procedures they would encounter and then teach each procedure using independent

W. MICHAEL REED is Professor, School of Education, New York University, 305 East Building, 239 Greene Street, New York, NY 10003-4721. E-mail: w.michael.reed@nyu.edu

[Haworth co-indexing entry note]: "Assessing the Impact of Computer-Based Learning: A Research Course for Graduate Instructional Technology Programs." Reed, W. Michael. Co-published simultaneously in *Computers in the Schools* (The Haworth Press, Inc.) Vol. 16, No. 1, 1999, pp. 151-171; and: *Information Technology in Educational Research and Statistics* (ed: Leping Liu, D. LaMont Johnson, and Cleborne D. Maddux) The Haworth Press, Inc., 1999, pp. 151-171. Single or multiple copies of this article are available for a fee from The Haworth Document Delivery Service [1-800-342-9678, 9:00 a.m. - 5:00 p.m. (EST). E-mail address: getinfo@haworthpressinc.com].

variable(s) and a dependent variable related to instructional technology designs.

Given the timing of my concern–that my students were approaching the point of developing their thesis or dissertation designs–I included as one of the major tasks their drafting a research proposal with an introduction, a review of literature, a rationale, research questions, and a design of the study in which they explained their sample, independent and dependent variables, procedures, and analysis of the data. They would also include a reference section and appendices, if needed for detailed, lengthy description of the treatment, copies of any instruments they plan to use, and other information not neatly included in the other sections. Besides this serving as an academic task to bring together what they had learned, for many of them the document served as an early draft of their thesis or dissertation research proposals.

There is always the issue of courses overlapping in content. It is true that the students taking this course have already had some of the information or that students taking this course early will face much of the information later in other courses. But, given the concern I had from which this course was a result, it is probably necessary that they are re-exposed to some of the same information. The course is more than just a statistics course, so there are other reasons for offering it. Re-visiting the information and then in a personally meaningful context can only be of help to them. The more we can expose them to it, the better prepared they will be.

In this course, the students are exposed to common readings, they also select six articles of their own choosing to critique, they learn a statistics program that they will have easy access to throughout their graduate programs, and they are required to display their understanding of the course's content via the statistics test and the final research project.

Appendix A is a week-by-week account of the course. Appendix B is the critique guidelines the students follow when critiquing the articles they have selected. Appendix C is a sample statistical exercise used during the class meetings. Appendix D is a sample test item. Appendix E is the guidelines for developing their research proposals. Each evaluative activity is explained under the next section I have called Strands.

STRANDS

Common Readings and Response Logs

The common readings are those that all students read. These are readings that I feel are important for them to know. So far I have made sure that they read Kozma's review on media studies (Kozma, 1991), Clark's work on media research (Clark, 1983), Palumbo's review on problem-solving and computing (Palumbo, 1990), and a special issue of the *Journal of Research on Computing in Education* that I, Dennis Spuck, and William Bozeman (1996) edited entitled *Assessing the Impact of Computer-Based Learning*, which contains reviews of research on computer use in writing, science, mathematics, social studies, hypermedia, and special education. Besides the content, these works serve as excellent examples of reviews of literature. I then select research articles that feature some of the most prominent independent variables and dependent measures. Because they are also exposed to the eight primary statistical procedures on a weekly basis, I try also to line up a research article that includes the statistics procedure for that week (see Appendix A for a week-by-week account of the course).

The students are required to submit a response log based on the weekly readings. Lately I have been requiring them to E-mail me their response logs a few days before the course meets. In the response logs, they are to identify five to eight points that they found especially interesting or confusing and to state why. Getting the logs prior to the class meeting allows me to read through them and identify topics we may need to spend more time on given their confusion. Sometimes they ask for more information on a topic; it gives me enough time to prepare some statements on them so that I can address them in class. I react to them informally, in the spirit of journal writing. I do not grade them per se or assess their writing performance or ability.

Usually half of each weekly meeting is spent discussing the readings. The response logs also assure that they have read the articles and then serve as a cueing device when we have open discussions. I return the response logs for them to refer to during this part of class.

Student-Selected Research Articles for Critiquing

In addition to the common readings, each student selects and critiques six data-driven articles (see Appendix B for the critique guidelines). If they do not already know how to access ERIC online, I give them the URL for that. For the second class meeting, they are to turn in a list of 12 articles. This allows me to assure that they are data-driven articles and not reviews of literature or conceptual or applied pieces. Also, I want to assure that all students are critiquing different articles so that they can collectively be critical of as many different articles as possible. They are required to briefly present two of their six articles to the class, pointing out the research questions, the design components, the results, and their assessment of the quality of the two articles. Depending on the number of students in the class, how many students provide a critique each class meeting will vary. In a class of 18 students, there will be six per class meeting over a six-week period. It is important to limit their oral critiques to five to ten minutes, given limited class time and much else to cover.

In addition to their orally presenting a critique of two of their articles, I have the students reduce each critique to a two-page paper, single-spaced, and, if needed, smaller font, near the end of the course as well as a bibliographical listing of their six articles and turn in copies to me. I then put the revised, reduced critiques on reserve for them to read or copy later. I make enough copies of the listings of the articles and distribute those in class. By making such information available and each student critiquing six articles different from those critiqued by the rest of the students, they are helping one another identify articles and their relative usefulness and quality. This may help them with their proposals or later research. In a class of 18 students, that means that 108 articles will have been critiqued and can easily serve as an important resource as they identify studies relevant to their research. These 108 articles are in addition to the numerous research and review articles they have read as common readings.

Research Text

Any research text will probably suffice. It should cover general research terms and the eight primary statistical procedures covered in this course: paired *t*-test, unpaired *t*-test, oneway ANOVA, multi-factor ANOVA, ANOVA with a repeated measure, ANCOVA, simple regression, and multiple regression. Additional information covered in the course includes post hoc tests, reliability, validity, chi square, and creating subgroups of a sample using means and standard deviations on a logical grouping procedure.

Statistical Procedures

Each statistical procedure is covered in a given week; depending on school-related holidays and other interruptions, I sometimes combine a few, but they need to be highly related. For each statistical procedure, I generate a data set that they enter. A research context is first presented, followed by the research question. They enter the data, and then a series of questions direct them to certain statistical information. T-value and F-ratio tables are distributed when appropriate. They are then prompted to interpret the results. I usually include a follow-up exercise for them to do outside class. (See Appendix C for a sample exercise on the oneway ANOVA.) I have used Statview in this course, but any microcomputer-based package will work. To help anchor the research in a context meaningful to them, independent variables and dependent measures highly related to instructional technology research situations are used.

Prior to the test, I assign each of them to develop an exercise on one of the eight procedures. They then bring enough copies to class for distribution two weeks before the test. We spend the class meeting the week before the test going through a practice test, the in-class exercises, the follow-up exercises, and the student-created exercises. They will have then had four or five exposures to each of the eight procedures.

Statistical Test

The test is an open-book, open-notes, three-hour examination. It is one of the few courses I teach in which there is a test. The purpose of the test is not limited to understanding each procedure but rather is intended to simulate what the students will do as researchers: Given a research situation, what analyses will they employ? We, as researchers, do not engage in our research lives conducting just a series of paired *t*-tests or regressions. Rather, we are faced with a research context and need to identify which procedures we will use. The test is heavily context-based. A scenario is presented for each of the eight statistical procedures. The students have to analyze the features of the scenario, usually the number of independent variables and whether effects or relationships are being determined. In most situations, knowing that information will result in their successfully identifying which procedure to use. The questions for each procedure prompt them to identify the independent variable(s), the dependent measure, and whether effects or relationships are being posed. They are then prompted to identify which statistical procedure is needed (see Appendix D for a sample test item for the unpaired *t*-test).

For some of the scenarios, they enter data and analyze the data. For some, the statistical results are presented; and they are to explain the results. For some, besides identifying independent variable(s), dependent measure, and statistical procedure, results are presented, and they are to interpret the results.

What is required is to construct a test that makes them compare a scenario against the others and determine which procedure is to be used. There ought to be a fairly even distribution of scenarios that require them to enter data, to explain statistical results, and to interpret results. The test should be designed so that if all they had to do was proceed through the test in a very sequential, non-comparison/reflective manner, they could finish in one and one-half hours. Because they need time to compare scenarios and interpret, the additional one and one-half hours should provide ample time. In the 10-plus years I have taught this course, no one has ever finished in less than two and one-quarter hours; only a few of the 150-plus students have run out of time.

As I said, the test is open notes and open book. This is an attempt to simulate what they face as researchers: Given resources available to them, they make decisions about procedures, enter and analyze data, and interpret results. The only facet of the real-time researcher that is not allowed is collaboration; they must take the test by themselves.

Other Topics

Other topics include validity, reliability, a graphics software program for doing their figures (I use Cricket Graph), and drafting a grant proposal. I show them how to calculate the necessary information for the Cronbach alpha and the KR-20.

For the grant proposal, I choose a funding agency that will finance dissertations and provide them with the guidelines for them. Depending on the agency's requirement for the grant document, I usually tell them to wait until after they have written the research proposal. They will need to reduce parts to fit the usually very stringent length requirements of grant proposals; they will also need to create a timeline and a budget. But, the bulk of their work is eliminated if they wait until after writing the proposal and then work with it. Having some experience at securing funds is important. As financially challenged students, their being funded for even $500 makes it financially less difficult. Certainly having been funded is an important vita line and provides valuable experience in order to secure funds later on, when they have finished and have assumed their new positions.

Research Proposal

The culminating assignment is to draft a research proposal that contains a review of literature; a rationale; research questions; and the design components of sample, independent variables, dependent measures, procedures, and analysis of the data; references; and appendices (see Appendix E for proposal guidelines). For many of the students, the proposal does become their dissertation proposal. If not, it serves as the plan for a pilot study.

The next-to-the-last class meeting we do not meet formally as a class. Rather, students meet with me individually to discuss their drafts. I usually spread these meetings across the Tuesday, Wednesday, Thursday, and Friday of that week; with a class of 20 that will require five 30-minute meetings each day. They are to turn in their drafts by Monday afternoon, which allows me the night before to read the papers for the next day's meetings.

By having these meetings, it requires my students to write a serious draft weeks before it is due. Helpfully nudging them is always a plus. Also, I have found that the final products are usually better because of this individual attention. By having individual meetings, students can pose questions and ask for assistance that they may be less willing to bring up in class.

The last class meeting of the semester is dedicated to the students' presentation of their proposals to the class. The presentation class is set up much like a conference. They are required to develop a presentation using a program such as *PowerPoint*. I usually limit their presentations to 15 minutes. With 18 students and two papers during a concurrent session, 135 minutes are needed for the nine sessions. With a few minutes between each session, another 15 minutes are needed. With 180 minutes set aside for the three-hour course, 40 minutes are left over. Usually about halfway through them, I'll schedule the typical 15-minute break. That will allow me 10 minutes after the final session to go over closure items for the course and then dismiss class 15 minutes before the hour, which is the usual dismissal time. If there are fewer than 18 students, then the time limit can be increased, but a time limit similar to those at actual conferences should be employed. If there are more than 18, then more than two sessions can be given concurrently. Or, the instructor may want more bring-to-closure time and adjust the presentation schedule accordingly.

For the presentations I create a conference program, listing the presenter, presentation title, and a one-sentence explanation. Students then choose which of the two sessions they want to attend. I do this for a few reasons. One, it simulates presenting at a conference. Two, students are required to honor the 15-minute time limit which is designed to keep them from aimlessly presenting information (the next presenter setting up for his or her presentation is an obvious cue that time has elapsed). Third, we do not have to listen to all of them, which can become monotonously long and repetitive, despite how good they are. Fourth, students have to get up and move to the next presentation that maintains some of the energy for that evening.

I reserve two rooms near one another that have presentation systems in them (or portable systems can be brought in) so that a two-minute interval between sessions is sufficient time. Students have until the next week after the presentations to fine tune their proposals and should use any feedback from the student-audience as possible needs for change. I move between the two rooms, making sure presentations are starting and ending on time. It is not critical that I listen to all of each presentation, which, of course, would be impossible. The intent of this exercise is for them to go public with their studies and for the student-audience members to benefit from hearing about the other students' studies. I have read an early draft for the individual meetings and will be reading the final documents as I finalize grades. I will know their proposed research well enough without listening to all presentations in their entirety.

Some Additional Explanations of the Week-by-Week Account

Appendix A lists the week-by-week account of the course. Besides reviewing the requirements and purpose of the course in Meeting 1, I spend a little time reviewing some basic research terms. I explain that if students can determine how many independent variables there are, whether the design requires a pre/post comparison, and whether effects or relationships are being sought, they can significantly reduce the number of possibly appropriate procedures. It is not especially important that they understand all of that so early in the course. But, they should sense that such considerations are at the root of selecting statistical procedures. By the end of the course, they should be able to follow that process.

Common readings dominate Meetings 2 through 9. These consist of articles, both reviews and research reports, that I feel students should know as instructional technology researchers. As I said previously, I also try to match an article for common reading that features the statistical procedure for that week.

During Meeting 2, they turn in their list of 12 research articles; I return the list with the approved articles marked during Meeting 3. An article may not be approved because someone else has listed it or because it is not a research article, but rather a theoretical or applied piece or, in some cases, a meta-analysis.

In Meeting 2, I use the statistical exercise of determining the mean and standard deviation for subgrouping the sample for a variety of reasons. First, conceptually it is very complex so that allows students to deal with the procedural "mechanics" of the *Statview* software. Second, I think such a procedure is important because more and more classroom-based research is being conducted where researchers are handed intact groups. One way to make the design more complex is to subgroup the sample on a logically related measure and see how that measure or factor interacts with the treatment.

As can be seen for Meeting 3, I have combined the unpaired *t*-test and the oneway ANOVA, due to the time demands of the course. These two procedures are closely related because the unpaired *t*-test deals with two groups, or two levels of one independent variable; and the oneway ANOVA deals with more than two groups, or more than two levels of one independent variable. This combination is an effective learning moment to discuss post hoc tests. While any significant *t*-value for the unpaired *t*-test clearly points to the two means of the two groups being statistically different, the significant *F*-ratio for the oneway ANOVA does not do that for the three or more means; rather, the *F*-ratio simply signifies that some of the means, perhaps all, differ from one another. The post hoc will compare mean pairs and identify which means actually differ.

Likewise, I have combined the simple and multiple regressions in Meeting 8. Regression information such as beta-weights has no meaning unless there is more than one independent variable. It is also interesting to see what happens to the R-squared when there is more than one independent variable. Because different independent variables will have different ranges, discussing beta, or the slope, for each independent variable helps anchor what beta means.

Meetings 4 through 9 feature the oral critiques as well as when the written critiques are due. Early in the course, I allow students to leave blank some of the critique components because we may not have gone over them yet. I copy those critiques for my records. When the students feel they can address those components, they then turn in the more complete critiques. If they never do turn them in, I grade the earlier critique.

Statistical procedures are features of Meetings 2 through 8. For homework due at Meeting 9, I assign each student to develop a practice exercise on one of the eight procedures. They are to also test out their exercise and provide an answer sheet. During Meeting 9, they bring enough copies to hand out to the students and me.

Meeting 10 is set aside for reviewing the practice test, the class exercises, the follow-up exercises I provided, and the student-created exercises. They are to come to class having done all of them, with very specific questions they need to pose.

In preparation for the test during Meeting 11, I tell them to review everything but just as importantly to organize their notes, the exercises, and other information they want to use during the open-book, open-notes, open-exercise test. They should minimize the time for searching for information during the test.

The activities for Meeting 12 are mostly explained earlier in this article. When they finish the test during Meeting 11, I assign them chapters on validity, reliability, and chi-square. Even though I believe chi-square is a statistical lightweight, it is a procedure they see often in the research. I also give them a sample grant proposal for them to become familiar with. They are also to reduce their critiques to two pages each and turn in clean copies to me; I then put them on reserve for their later use. They are to bring in enough copies of the one sheet on which is listed the bibliographic information for the six articles they critique. These lists are handed out. A final task I have them do is to list five likes or dislikes of the research articles they read and to explain why they are likes or dislikes. They are to bring in enough copies of these lists to hand out to the rest of the students and to me.

The purpose behind this is that usually the likes and dislikes are based on how studies were written up and/or presented. For example, they might like a study because the results were clearly displayed in table form. Or, they might dislike a study because the measurements were not completely explained or there was some missing information they expected when the researchers described the sample. After hearing a like and/or a dislike from each or most of the students, depending on the number of students and the time allotted, I then point out to them what has by that time become the obvious: Do not do what you disliked, but do what you liked, in your own research.

Ending the course are the individual meetings to discuss their proposal drafts and then the presentations. I set aside exam week for students to polish up their research proposals and complete the grant proposals.

Limitations of the Course

One of the major limitations of the course is that it is quantitative. Some of the quantitative portions could be eliminated and qualitative information could be inserted. I generally advise against this because covering just one of the two primary methodologies is challenge enough for most students. Plus, it requires that the instructor be amply well versed in both, which is a rare find. Another solution is to design a separate qualitative course or to collaborate with the qualitative expert and have him or her meet with the students to discuss some possible connections with qualitative methods and instructional technology.

Another limitation is that, except for chi square, the course does not include non-parametrics procedures. Also, there are other parametric procedures not covered, including procedures in which more than one dependent variable is used.

Despite these limitations, there is the driving issue of time and deciding on the information most meaningful. Other courses can be created to address the aforementioned limitations.

SUMMARY

The course has proven to be very successful. I find myself spending much less time informally "teaching" students one-to-one than during the early days. More importantly is that they appear to be much more confident in themselves as researchers. For many, the research proposal moves them closer to the dissertation proposal.

REFERENCES

Clark, R. E. (1983). Reconsidering research on learning from media. *Review of Educational Research, 57*(4), 445-460.

Kozma, R. B. (1991). Learning with media. *Review of Educational Research, 61*(2), 179-211.

Palumbo, D. B. (1990). Programming language/problem-solving research: A review of relevant issues. *Review of Educational Research, 60*(1), 65-89.

Reed, W. M., Spuck, D. W., & Bozeman, W. C. (1996). Assessing the impact of computers on learning: 1987-1995. *Journal of Research on Computing in Education 28*(4).

APPENDIX A

Schedule of Class Meetings

Meeting 1: introduction to class
review basic research terms
review eight statistical procedures
have students articulate their research interests

Meeting 2: common reading discussion based in part on e-mailed reponse logs
turn in list of 12 articles
statistical exercise on using mean and standard deviation for subgrouping sample

Meeting 3: common reading discussion based on response logs
return list of approved articles
statistical exercises on upaired *t*-test and oneway ANOVA

Meeting 4: common reading discussion
critique #1 due
oral critiques
statistical exercise on multi-factor ANOVA

Meeting 5: common reading discussion
critique #2 due
oral critiques
statistical exercise on paired *t*-test

Meeting 6: common reading discussion
critique #3 due
oral critiques
statistical exercise on ANOVA with repeated measure

Meeting 7: common reading discussion
critique #4 due
oral critiques
statistical exercise on ANCOVA

Meeting 8: common reading discussion
critique #5 due
oral critiques
statistical exercise on simple and multiple regression

Meeting 9: common reading discussion
critique #6 due
oral critiques
distribute student-created statistical exercises
distribute the practice test

Meeting 10: review for test, using class exercises, follow-up exercises, and student-created statistical exercises

Meeting 11: test

Meeting 12: go over test
validity
reliability
graphics program
chi square
grant proposal format
distribute article likes and dislikes
distribute bibliographical list of articles
turn in reduced critiques

Meeting 13: individual meetings on research proposals, no class

Meeting 14: presentations, using a conference format
closure items discussed

Meeting 15: usually exam week, no meeting, turn in research proposals and grant proposals

APPENDIX B

Guidelines for Critiquing Research Articles

Section One

Bibliographic information (author[s], year of publication, title of article, name of journal, volume, issue, page numbers)

Purpose of the study

Research questions

Sample

Dependent measures

Independent variables

Analysis of the data

Results

Discussion

Section Two

(answer in detail; explain why yes or no is appropriate)

Is the problem important practically and/or theoretically?

Is the design appropriate for the problem?

Is the sample appropriate? Well chosen? Carefully selected?

Are the measuring instruments valid, reliable, and appropriate?

Is the report clear and complete? Is rigor evident?

Is the analysis appropriate and well thought out?

Are any generalizations made in harmony with the results? Are conclusions justified?

Could this research be replicated on the basis of what is reported?

APPENDIX C

Sample Statistical Exercise for Oneway ANOVA

SITUATION:

You are a math teacher, faced with the task of choosing mathematics problem-solving software. A company has sent you several pieces of software, all of which, according to the salesperson, promote mathematics problem-solving. You are not quite sure that each does. Because you have a limited budget and these software packages are somewhat expensive, you feel you need to test them out empirically.

You decide to try out the software designed for ninth graders; you've chosen one package that is predominantly drill-and-practice, another that is predominantly tutorial, and a third that is predominantly simulation.

You want to determine which type of mathematics problem-solving software results in the greatest mathematics problem-solving. Thus, your research question is

WHAT IS THE EFFECT OF TYPE OF MATHEMATICS PROBLEM-SOLVING SOFTWARE ON MATHEMATICS PROBLEM-SOLVING?

Because you have only one independent variable–TYPE OF SOFTWARE–and are interested in mean differences across type of software, you need to conduct a one-way analysis of variance.

You have administered a mathematics problem-solving instrument (FORM A of that test) prior to the students going through the software and have found that there are no statistically significant differences across the student-participants randomly assigned to treatment (or type of software). Thus, you feel confident that you can compare posttest scores, or Form B of the same instrument, to determine which type has the greatest effect.

On the next page are posttest scores, or scores resulting from their taking Form B of the same instrument.

Step One:

After turning on the Macintosh, double click the Statview 4.02 icon. Another menu will appear; double click the new Statview 4.02 icon that appears.

Click the full-screen Statview logo when it appears.

Under the banner at the top of the screen, click File (be sure to keep the button pressed down) and drag the arrow to New.

A spreadsheet appears.

APPENDIX C (continued)

Highlight the phrase **Input Column**. Type in the term SOFTWARE.

Move cursor to Real (to the right of Type), click, keep the button down, and move down to Category.

A window appears asking you to choose the column's category.

Select **New**.

In the second "box" (the one highlighted), type in DP and click ADD.

In the same box, type in TUT and click ADD.

In the same box, type in SIM and click ADD.

CLICK **DONE**.

Highlight the next phrase **Input Column** in the next Column. Type in the term SCORES.

Move cursor to Real (to the right of Type), click, keep the button down, and move up to Integer.

Click the gray box at the bottom of the first column you created. It should turn black.

Type in the term DP six times, hitting the Return key after each one.

You might note that once the computer can distinguish one category from the other(s), it will type in the rest of the word.

In the same column, type in the term TUT six times.

In the same column, type in the term SIM six times.

Click the first box of your second column. It should turn black.

Type in the 18 scores, hitting the Return key after each one. Make sure that a score corresponds with its correct SOFTWARE subgroup–that is, next to the term DP in the first column you created one of the DP scores is in the next column to the right of the term.

Drill/Practice	Tutorial	Simulation
20	43	34
22	31	26
25	38	32
17	17	27
12	49	41
23	41	33

After typing the in scores, click **New View** under Analyze.

Click the side-triangle next to ANOVA once.

Highlight ANOVA TABLE and MEANS TABLE. The triangle turns down.

Click SOFTWARE once.

Click Add once.

Click SCORES once.

Click Add once.

Click Create Analysis once.

Make sure that FACTORIAL has been marked.

Click Scheffe.

Click OK.

A Results Table appears.

Step Two:

What is the *F*-ratio?

What are the degrees of freedom (SOFTWARE, RESIDUAL)?

What must be F-ratio be for there to be significant mean differences (SOFTWARE across the top and RESIDUAL along the left side)?

Are there significant differences?

APPENDIX C (continued)

Step Three:

Based on this analysis, you are assured that there are mean differences, or at least that one type of software has a greater effect on mathematics problem solving than one or both of the others.

But, do you know which one?

What is the mean for DP?

What is the mean for Tut?

What is the mean for Sim?

You can look at the data and speculate. Clearly the difference may be between drill-and-practice and tutorial. But, is there also a difference between tutorial and simulation?

It is difficult to tell.

Thus, a post hoc test is needed. This procedure will let you know which means actually differ.

Which means actually differ?

Based on this analysis, what instructional decisions might you make about the purchasing of the software?

FOLLOW-UP EXERCISE:

Your assistant principal, a person who has had five disastrous experiences with computers, isn't impressed that any of the software has produced performance improvement. Being the self-centered person that she is, she stipulates that before "good money is spent on something that makes everybody EXTREMELY NERVOUS!" there must be proof that computer anxiety is lowered (SIGNIFICANTLY SO!).

Having had many exchanges with the lovely, yet-holder-of-the-departmental-purse-strings-as-assistant-principal, Ms. Ima Skeptic, you know it's futile to try to convince her without conducting another study.

You decide to use another ninth-grade math class. It won't be a waste of time because you can determine whether or not the results of your first study will hold true–rather than due to chance. You have grouped this ninth-grade class into three groups with six students in each group, have administered the computer anxiety instrument at the onset, and have found the students' computer anxiety is the same across TYPE OF SOFTWARE. You know that you can compare post-instruction computer anxiety scores to provide Ms. S the information she has demanded.

What is your research question?

Why is the oneway the appropriate procedure?

The post-instruction computer anxiety scores are as follow:

Drill/Practice	**Tutorial**	**Simulation**
45	30	70
33	27	60
28	33	55
40	20	40
37	27	47
41	30	62

Go through Steps One, Two, and Three (given previously) and answer the questions in relation to the new question.

Do your findings give you the support Ms. S is demanding prior to purchasing? Or, does Ms. S have evidence for her less-than-nice attitude about life in general?

If you wanted to determine whether the computer anxiety of one of the SOFTWARE TYPE groups significantly changed due to the software, what information would you need and what procedure would you use?

By the way, performance measures were consistent with Study One's findings. Something else you can throw in Ima's disappointed and disapproving face.

APPENDIX D

Sample Test Item for Unpaired T-test

Below is listed a data set of writing apprehension post-treatment scores for two groups of 12th-grade students in an Advanced Placement English class. One group's writing instruction allowed them to word-process their papers; the other group's did not. It was determined that the two groups' writing apprehension was not significantly different at the pre-treatment point.

The higher the writing apprehension score, the greater the writing apprehension. The instrument range is 26 to 130. Students responded to the instrument at the beginning of the semester and at the end of the semester.

Word-Processing Group	Non-Word-Processing Group
33	37
28	26
26	30
40	35
42	35
28	33
35	37
38	36
50	45
33	28

Enter the data using *Statview.*

What is an appropriate research question?

What is/are the independent variable(s)?

What is the dependent measure?

What statistical procedure is needed?

Why?

What is the *t*-value or *F*-ratio? What is/are the degree(s) of freedom?

Is word-processing more effective at reducing writing apprehension than not including word-processing?

In the context of the sample and the treatment, interpret the results?

APPENDIX E

Guidelines for Research Proposals

Introduction (create this last; it is meant to set the stage for the reader; explain the purpose of the study and the research context; aim for two or three pages)

Review of Literature (should be domain driven, rather than author driven; use headings and subheadings effectively to help direct the reader to the different facets of the research; studies and theories should relate directly to the research questions; aim for 12 to 15 pages)

Research Questions (what questions are you answering via conducting this study)

Design of the Study

Sample (explain the sample in terms of age and ability and how selected and via other information that may have some bearing on the study)

Independent Variable(s) (should include treatment as well as blocking variables; explain instruments used to group the sample, including validity and reliability information; make sure they appear in the research questions and there is reference to them in the Review of Literature)

Dependent Measures (explain instruments or tests, including reliability and validity information; these scores will indicate changes in performance or attitudes; they too should appear in the research questions and should be referenced in the Review of Literature)

Analysis of the Data (state the research question, the statistical procedure to be used to analyze the data, the independent variable[s], and the dependent measure)

Procedures (provide a step-by-step account of data collection and treatment implementation–do not repeat the context specifics of the treatment which you should have done in the Independent Variable[s] section)

References (all articles referred to in the proposal)

Appendices (any instruments used in the study; more detailed information about the treatment; other information relevant to the study that does not fit neatly elsewhere; be sure to reference each appendix in the text of the document)

TECHNOLOGY AND STATISTICS

George C. J. Fernandez
Leping Liu

A Technology-Based Teaching Model That Stimulates Statistics Learning

This study presents a technology-based statistics teaching model developed from five-year's experience in teaching statistics. Technology-based learning and cooperative learning are merged into this model, including practical teaching processes that stimulate student enjoyment and motivation to learn statistics.

LITERATURE REVIEW

Enrollment in statistics courses in U.S. colleges and universities has increased from 10% in 1966 to 52% in 1990 (Cobb, 1993). In a

GEORGE C. J. FERNANDEZ is Associate Professor, Department of Applied Economics, University of Nevada-Reno, Reno, NV 89557. E-mail: gcf@unr.edu
LEPING LIU is Assistant Professor, Department of Reading, Special Education and Instructional Technology, College of Education, Towson University, Towson, MD 21252.

[Haworth co-indexing entry note]: "A Technology-Based Teaching Model That Stimulates Statistics Learning." Fernandez, George C. J., and Leping Liu. Co-published simultaneously in *Computers in the Schools* (The Haworth Press, Inc.) Vol. 16, No. 1, 1999, pp. 173-191; and: *Information Technology in Educational Research and Statistics* (ed: Leping Liu, D. LaMont Johnson, and Cleborne D. Maddux) The Haworth Press, Inc., 1999, pp. 173-191. Single or multiple copies of this article are available for a fee from The Haworth Document Delivery Service [1-800-342-9678, 9:00 a.m. - 5:00 p.m. (EST). E-mail address: getinfo@haworthpressinc.com].

national survey, 39% of universities offered two undergraduate statistics courses, 35% offered one course, 6% offered three courses, and 18% offered four or more courses (Tabatabai & Gamble, 1997). Introductory statistics courses are especially important for education and social science majors because often these courses may provide the only exposure students have to statistical applications. However, students frequently view statistics as their most difficult and boring college course because the statistics teacher fails to create enthusiasm for and motivate students to learn statistics (Hogg, 1990). Therefore, methods for improving statistics education have currently become an active area of research. A review of literature indicated that improving statistics education is a dynamic, progressive, and sometimes controversial subject (Cobb, 1993; West & Ogden, 1998; Zetterqvist, 1997; Hogg, 1990,1991; Smith, 1998). One critical issue in improving statistics teaching suggested in this review is the failure of traditional statistics teaching methods.

Traditional methods of teaching statistics are often viewed as ineffective for several reasons. This study focuses on two. First, instructors have taken the fun out of learning statistics (Yilmaz, 1996). The traditional mixture of lecture and arithmetic problems tends to be too abstract, forcing students to devote too much time in the actual calculating process and therefore to lack an understanding of the overall structure about what they are learning or what a specific calculation is for. Second, students are not highly motivated to learn statistics because courses fail to establish a clear link between statistics and its usefulness (Yilmaz, 1996). Data sets used in teaching are often artificial lists of numbers created only to give computational practice. Students therefore see no practical application of their work. Sometimes, course materials are similar to a reduced version of a more theoretical statistics course, which provides neither the systematic theoretical background of a certain statistics area nor sufficient knowledge or skills for performing specific applications in that area.

Enjoyment, Motivation, and Statistics Learning

Enjoyment and motivation are key factors to enhancing statistics education because, as attitude variables, they have been found to be very important in influencing student learning and achievement (Liu & Johnson, 1998; Liu & Fernandez, 1998; Dietz, 1998).

Enjoyment as an attitude variable refers to how much students enjoy or are interested in learning (Liu & Johnson, 1998). It is assumed that, if one enjoys something, an effort will be made to learn it. The more effort one makes, the more one learns. *Motivation* can be defined as a willingness or desire to attempt to do something (Kellenberger, 1996). Motivation theories and models attempt to explain future behavior (Keller, 1983) and assume a relationship between motivation and performance (Locke & Latham, 1990,1994; Benswanger, 1991). The research also suggests that (a) these two variables–enjoyment and motivation–have linear relationships with learning achievement, and (b) the more students enjoy learning and are motivated to learn, the greater their learning achievement will be (Liu & Johnson, 1998; Liu & Fernandez, 1998; Garfield, 1993, Dietz, 1998).

Theoretically, as shown in Figure 1, these findings are applicable to statistics learning and teaching. That is, if teaching methods can make the statistics class enjoyable to students and can motivate students to learn, not only can we expect higher achievement to occur but also that students will *enjoy* learning statistics and be more *motivated to learn*. Therefore, based on research findings, enhancing my statistics teaching started with a reform of traditional teaching methods and content with the primary goal of stimulating learning.

Teaching Methods That Stimulate Statistics Learning

Research on enhancing statistics education has focused on three approaches: (a) cooperative learning, (b) activity-based learning, and (c) technology-assisted learning. When these three approaches are integrated into the teaching/learning process, students are stimulated to learn statistics and consequently effective learning occurs.

Cooperative Learning. Cooperative learning gradually has been used in statistics classes to supplement or replace the traditional lec-

FIGURE 1
Factors related to enhancing statistics education.

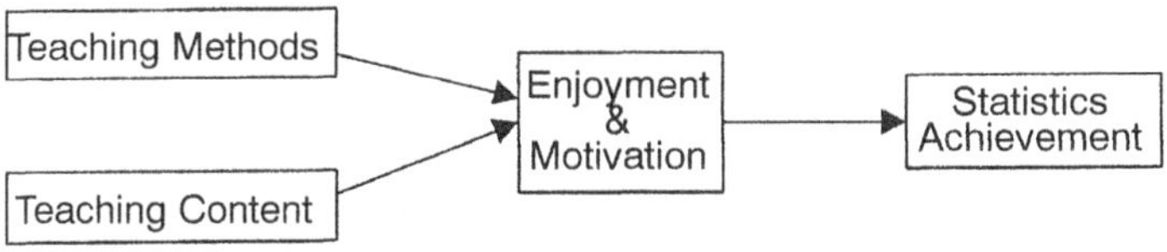

ture (Gnanadesikan, Scheaffer, Watkins, & Witmer, 1997; Keeler & Steinhorst, 1995; Garfield, 1993). The purpose of these learning activities is to actively involve students in learning and to develop improved skills in working with others (Garfield, 1993). In the literature, the most frequently recommended cooperative learning activity is group or team projects.

In group projects, two or three students work together to solve a problem, discuss a procedure, or analyze a set of data. Group projects also provide a means to help students develop statistical reasoning (Smith, 1988) by incorporating cooperative-learning strategies to supplement what students have heard and read about statistics: designing studies, collecting data, analyzing results, preparing written reports, and giving oral presentations. Hunter (1997) and Roberts (1992) describe two successful implementations of group project activities for engineering and MBA students. Research by Garfield (1993) and Dietz (1998) has shown that student motivation and enjoyment can be enhanced by group projects.

Activity Based Learning. Hands-on activities to illustrate basic statistics concepts promotes the teaching of statistics as more of an experimental science and less of a traditional mathematics course and enhances learning by improving student attention, motivation, and understanding (Gnanadesikan, Scheaffer, Watkins, & Witmer, 1997).

Technology-Assisted Learning. Another strategy for enhancing statistics education is to integrate technology into the teaching/learning process (Webster, 1991; Groeneboom & de Jong, 1996; Gatti & Harwell, 1998; Christmann & Badgett, 1997). A 1991 survey indicated that 70% of introductory statistics courses used computer software (Khamis, 1991). Initially, data analyses were performed by writing program code with statistics programming software. However, with the introduction of the PC/Windows environment in the early 1990s, menu-driven interface became the standard form of statistics software (Fernandez, 1988). Currently, programming applications and point-click menu-driven applications are the two main categories of statistics software.

The program-code approach is often considered a more powerful and flexible method of data analysis (Fernandez, 1998). Some program-code statistics software packages, such as SAS and the initial version of SPSS, have been widely used in industrial fields and, to a certain extent, in education. However, even though this approach is very powerful, it is not considered user-friendly and students need to have some programming experience. Most of the student's time is spent on writing program code and debugging program errors. Therefore, the extent to which the program-code approach can be taught to average users and introductory statistics students is limited (Fernandez, 1998).

In contrast, statistics software in the Windows environment, with menu-driven interface, is considered user-friendly and easier for beginners to learn. Some popular graphics software companies, responding to user needs, have developed dynamic, interactive, and user-friendly data analysis modules. These modules have become more and more popular among data analysts and are accepted as the industry standard. However, to what extent they can be adapted for classroom teaching/learning requires further exploration. Some menu-driven statistics software packages in the Windows environment, such as Minitab, Easystat, and Asystat, are not considered as powerful as the program-code statistics software; they have limited capabilities to perform complex data analysis or advanced statistics (Fernandez, 1998).

Both types of statistics software have their advantages. Therefore, the question is: What approach should one use to teach introductory statistics and how should the specific software be integrated into the teaching/learning process? And, since the research shows that students are highly motivated when they learn with computer technologies (Astleitner & Keller, 1995; Malouf, 1988; Brush, 1997), another important question is, How should a statistics teacher design a technology-based course to motivate learning?

Technology-Based Cooperative Learning. Since research findings show that cooperative learning, hands-on activities, and technology-assisted learning can motivate student statistics learning and produce higher learning outcomes, it is proposed that a combination of technology-based cooperative learning with hands-on activities is an effective way to teach statistics. Alavi (1994) examined the learning outcomes of technology-based cooperative learning and reported positive results. There are many ways to integrate technology into statistics teaching and into cooperative learning (West & Ogden, 1998). One purpose of this study is to explore ways of adapting technology-based cooperative learning for the statistics class.

Teaching Content That Stimulates Statistics Learning

Summarized in the literature review, the content of statistics courses can reduce the extent to which students are motivated to learn. A survey of 500 U.S. higher learning institutions of introductory statistics course content showed that the most prevalent topics covered are data analysis, probability, sampling distributions, estimation, means and hypothesis testing. In upper-division statistics courses, regression analysis, hypothesis testing, time series, and forecasting are the most prevalent topics covered (Tabatabai & Gamble, 1997). Besides the techniques of data analysis and interpretation of output, one critical focus of teaching content in this study is the selection of data sets.

Fictitious Data. Data should be the foundation of the study of statistics. Cobb (1987) and Moore and Roberts (1989) have discussed the importance of introducing students to statistics through data-oriented courses. Students usually work with two types of data–fictitious or real. Fictitious data, for example, could be the arbitrary selection of a set of numbers (Khamis, 1991). Also, the instructor could use generated data that requires simple calculations (Posten, 1982; Read & Riley, 1983). However, studies show that students are not motivated in their learning when they work with fictitious data sets created to provide computational practice (Yilmaz, 1996). Students use calculators to do textbook exercises. They use the same data sets, do the same data analyses, interpret the same outputs, and write almost identical summaries or reports (Fernandez, 1998). Few opportunities are provided for students to make appropriate decisions on the choice of summary statistics and analyses, or to justify these decisions.

Real-World Data. Cobb (1987) suggests that using real data sets is an alternative method to improving statistics learning. A real data set has its purpose, cases, or variables in a real-world setting. When students know what the numbers mean, they became more involved in the learning (Hogg, 1972) and their data analyses become the search for meaning, not just exercises in arithmetic (Khamis, 1991). In using real data sets, the instructor can point out some real-world problems, such as potential difficulties in sampling, lack of certainty concerning distributional assumptions, and the complex processes of decision making. More and more statistics educators are teaching with real-world problems, projects, and data sets, which is believed to be an effective way to teach statistics and motivate learning (Hogg, 1972).

This literature review provided the research findings about, and the background for, using technology-assisted cooperative learning with real-world data to improve statistics education. My (senior author) five-years of experience in teaching statistics provided the practical processes from which a teaching model was developed to stimulate and motivate student learning.

REFORMING CLASSROOM TEACHING

The Problems

Using the traditional lecture-assignment method, I taught the statistics courses in the Applied Statistics Department of a western state university for several years–that is, until several problems became apparent:

1. Coming from different departments and diverse backgrounds, many students felt that the theories, concepts, and data analyses required were too abstract to learn, particularly those students from social science and education backgrounds.
2. Students apparently were not motivated to learn statistics.
3. Because it was necessary to condense statistics concepts and applications in order to cover the material within the allotted class time, there were too many concepts for students to fully understand; and, after about 50 minutes of class lecture, students began to "tune out," apparently unable to comprehend or concentrate further.
4. There was insufficient time in class for anything (e.g., quizzes or related activities) more than the lecture.

The Technology-Based Plan

With the objective to resolve these problems, I began to reform my classroom teaching by

1. Using cooperative learning instead of teacher-centered instruction
2. Using real-world data instead of fictitious data
3. Integrating technology into my teaching

Although each component was undertaken simultaneously in the reform process, technology became the axis around which the processes and activities of cooperative learning and the application of real-world data were designed and implemented.

Cooperative Learning

To deal with student diversity and to enhance student motivation, cooperative learning was adapted into both an introductory and an applied statistics course. The cooperative learning included student grouping, hands-on group activities, and group projects.

Student Grouping. Initially, two grouping strategies were used. First, students formed groups of two to three students according to their major area of study and interest. The advantage of this grouping, of course, was, with similar interests, it was easy to select a topic, design a study, collect data, perform a data analysis, and compose the report for the group project. For groups such as social science or education majors, who often lacked strong mathematics backgrounds, the instructor was able to provide specific help that other groups might not require.

The second strategy was assigned grouping by the instructor, who selected for each group someone who was good at computer technology, someone who had a strong mathematics background, and someone who had previous research experience. The advantage of this grouping, of course, was that students could help one another and learn from one another in a cooperative setting. The disadvantage was that, with diverse backgrounds, group members had difficulty focusing on similar interests.

Since self-grouping was more enjoyable to students and they seemed to be more motivated by this approach, eventually the instructor adopted self-grouping solely and provided additional help to groups with special needs.

Hands-On Computer Activities and Group Projects with Real World Data. Because students needed to use the SAS system in all the statistics courses, hands-on computer activities were conducted for the groups. Initially, students were given fictitious data to practice the basic functions of the software, and lab modules were assigned to each group as well as to each individual. From these lab modules, students gained basic knowledge about the SAS environment and confidence in using the software. Since students seemed to lack enthusiasm when using artificial data sets, after they learned the basic applications, they were provided with real-world data sets, some from ongoing research projects and others from sites on the World Wide Web. A completed activity, therefore, resulted in the students' finding an answer to an interesting question as opposed to simply coming up with a sterile number.

The initial objective of group projects with real-world data was to provide opportunities for students to acquire competence in carrying out common statistical procedures related to their professional area of interest. As a consequence, students were able to connect their conceptual knowledge in that area to relevant statistical knowledge. They were allowed to participate in projects relating to their current employment or their theses or dissertations. Using real world problems and data, the instructor had the opportunity to point out potential and practical difficulties of using statistics in the real world (e.g., sampling difficulties and lack of certainty regarding distributional assumptions). Students began to realize that very few real-world data sets were perfectly normally distributed.

Group projects were a semester long, and students learned a thorough statistics process and the necessary detailed techniques in each step–from analyzing data and checking assumptions to interpreting results and writing reports. They learned to use statistics software to perform data analyses. They saw the direct "products" of their work that could be used in their professional areas. It was this productive technology-oriented process and the practical subject-related outcome of group projects that motivated student learning and aroused student interest.

Another positive by-product from the group projects or activities was that students were more creative when they had the option to design their own projects and they showed less anxiety when working together. For those with weak computer technology or mathematics backgrounds, cooperative learning provided opportunity to discuss and learn from other experienced group members.

Integration of Technology

In order to meet our initial objective of integrating technology into statistics teaching to (a) effectively use class time by teaching within the time frame of students' ability to concentrate and (b) motivate students to enjoy learning statistics, different strategies were used:

Multi-processing presentations were used during the class lecture; SAS macros–an innovation in statistics education–were developed to help students quickly start using the software and effectively perform data analysis; technology workshop-oriented courses were offered to help students master technology skills; and information from Internet resources was used as complementary course material.

Multi-Processing Presentations. In the introductory statistics course, presentations consisted of explanations of concepts, design principles, and calculations; data sets demonstrations, using statistics software, and interpreting data analyses results. Most of these components needed to be exhibited, demonstrated, or explained simultaneously, and therefore the traditional class presentation of a linear sequence using the board or transparencies was inadequate. For example, during my discussion on using SAS data, I needed to show students the data set in a spreadsheet and the SAS code that converts the spreadsheet data into SAS data. And, during my discussion on interpreting data analysis results, I needed to show students the output and the plots consistent with the results. Since Microsoft Windows provides a multi-processing environment that allows several applications to be run simultaneously, I developed Word presentations to demonstrate concepts, theories, calculations, and report documentation; spreadsheets to demonstrate data sets or data charts; PowerPoint slides to demonstrate design models with color and animation; and SAS macros to demonstrate SAS code, results, and plots. With the Windows environment, I could open several applications simultaneously and switch the screen among all the applications.

The greatest benefit from using the multi-processing presentation method was effective use of class time. I was now able to present three times more information than with the traditional lecture method, and still there was adequate class time for discussion and group work. Students began to be excited and interested in learning statistics. Even when I increased my presentation time to 75 minutes when necessary, students continued to show great interest and were able to concentrate on learning.

Using SAS Macros. The SAS macros were an innovative approach to teaching statistics. Because SAS is such a powerful and flexible tool in performing advanced data analysis, I chose to use this software in my teaching. However, recognizing that it would be overwhelming for a beginner to learn both SAS programming and statistics in the limited class time, I developed several macros to help students get started and to explore the potential SAS applications according to their specific needs.

A macro includes all the program codes in one file that perform all the required functions of data analysis for certain statistics methods. For example, the macros for regression can perform: (a) assumption checking, including the tests for normality, unequal variance, and outlier, as well as the plots for these tests; (b) descriptive analysis of the data; (c) correlation analysis with the correlation coefficient matrices; (d) simple and multiple regression, including linear and nonlinear model detection, regression coefficients, significance of each variable with partial regression plots, significance of terms with all combinations of the variables (e.g., quadratic terms and cross-product terms); (e) robust adjustment with all the necessary results and plots; (f) logistic regression with the necessary results and plots; and so on. With all of these functions available to them on a macro, students are able to perform the statistics without first learning the complex programming, thus focusing their study on statistics, not the complex codes.

When students use the SAS macros, they simply learn how to use the "call file" to call certain functions in the macro to perform the data analysis. For example, to perform multiple regression analysis, they use the call file, as shown in Table 1, to call the macro "MULTLREG. MAC" to perform all the necessary tests of multiple regression. In this call file, the file name of the data set is "fat," the response variable is Y, and the predictor variables are X1, X2, and X3; the quadratic terms to be tested are X1sq, X2sq, and X3sq; the cross-product terms to be tested are X1X2, X1X3, and X2X3. The resultant output is displayed in the "Output" window of SAS, and students can save the results onto a floppy disk. All the graphics produced from the macro are displayed in the "Graphics" window, which also can later be saved on disk.

Using the macros, students are able to quickly start "doing" statistics with real-world projects. As they learn additional statistics concepts, rules, or applications, it is easier for them to understand the codes, the functions some specific codes can perform, and the results or graphics the specific codes can produce. They can see the concrete "products" of their work, and they become very excited about their

TABLE 1
Sample Call File for Multiple Regression Analysis

```
*name of this file is a:\multlreg.sas';
%inc 'a:\macro\multlreg.mac';
%multlreg(data = fat                  , /*RQ: SAS data set name */
          resp = y                     , /*RQ: Response variable */
          pred = x1 x2 x3              , /*RQ: Predictor variables */
          term = x1 x2 x3              , /*RQ: model terms */
        weight = %str(          )      , /*WLS option: Weight _wt_ ; */
         other =                       , /* Other model options: noint */
          test = %str(          )      , /* Parameter test option */
      restrict = %str(          )      , /* Restricted option */
            ci = 95                    , /* Level of Confidence */
       allcomb = x1sq x2sq x3sq x1x2 x1x3 x2x3          ,
                                         /* B_P test terms */
          size = 1                     , /* Font Size of title and axis */
           dir = a:\plot               , /* Directory to save graphs */
           dev = win                   ) /* Change to "CGM" to save your graphs */
       *RQ: Required values ;
```

work and are willing to do more and learn more. The most boring course becomes interesting and useful.

Workshop-Oriented Course. To effectively integrate technology into statistics teaching/learning, students are prepared with the necessary computer technology and skills for using the statistics software by participating in a workshop-oriented lab course, running parallel with the technology-based statistics courses. Working with lab modules I developed, students gain basic computer skills and experience with using the statistics software. Hands-on experiences are provided for (a) working with several applications in a multi-processing environment, (b) copying files between applications (e.g., copying SAS output results to a Word file), (c) managing data from different sources (e.g., converting data in a spreadsheet or a database into SAS data), (d) performing descriptive analyses with SAS, and (e) simple SAS programming.

It is recommended that students with little computer experience take this workshop-oriented lab course. With the same instructor of the statistics courses teaching this lab course, students are less anxious in their learning. In addition to learning computer skills, students have opportunity to ask questions related to both courses. After only a few weeks of participating in the lab class, students exhibit more and more confidence with using computer technology and SAS software. What was initially an experiment, this lab course is now considered a very effective and vital component of my statistics teaching.

Using Internet Resources. Another component integrated into my technology-based teaching is using Internet resources. With the rapid development of Internet applications, especially World Wide Web (WWW), rich information from various Web sites was integrated into my statistics courses from both an instructor and student perspective. First, I adapted Internet information using the latest WWW technology into my course interactive demonstrations in the form of Java applets. A number of WWW locations provide information for statistics education in electronic form (see the Appendix). Second, students use Internet information as complementary material for the course (e.g., some Web sites [see the Appendix] provide valuable real-life data sets from which students can select appropriate data for their learning). Information from the Internet for both the instructor and students goes far beyond what any textbook can provide and enhances student learning by providing a multi-dimensional information network for the course.

CONCLUSION

My initial purpose for reforming my statistics teaching was to motivate students to learn and therefore to improve their statistics achievement. During a five-year period, cooperative learning and technology-based teaching were gradually integrated into my statistics courses. The teaching strategies I developed provided, to a certain extent, solutions to the problems I observed while using traditional teaching methods (as outlined earlier in this article). Four major outcomes demonstrate an improvement in statistics teaching/learning. First, cooperative learning and technology-based teaching motivated students with different backgrounds to learn statistics. Through active participation students demonstrated increased interest and confidence in learning statistics. Some of their comments about my courses follow:

- *The project was a good way to tie the course together.*
- *The deadlines were reasonable and I actually enjoyed seeing the project go from its early stages to the final completed report.*
- *I really enjoyed the course. The use of the computer presentations was excellent. Using this format for our final presentation was great.*

- *The project was interesting because it was a topic of our choice.*
- *I am an education major and I really want to take more statistics courses like this.*

These comments reflect students' reduced anxiety and increased motivation for learning statistics.

Second, with technology, I was able to effectively use the limited class time, presenting much more information to the class than with the traditional teaching method, and students were able to concentrate longer and with more interest.

Third, student enrollment increased from about 28 students in 1993 to nearly 50 in 1998 (see Figure 2).

Fourth, students' average scores in the statistics courses increased during the five-year reform period (see Figure 3). Before integrating technology (1995), average scores were between 75 and 78; whereas, following the integration of technology, average scores increased to between 85 and 87 (1998).

The improvements shown in Figures 2 and 3 exhibit the changing situation during this five-year period. However, because this was not an experimental control study, we cannot interpret these improvements as the "effects" of teaching strategies. These outcomes do,

FIGURE 2
Introductory statistics course student enrollment.

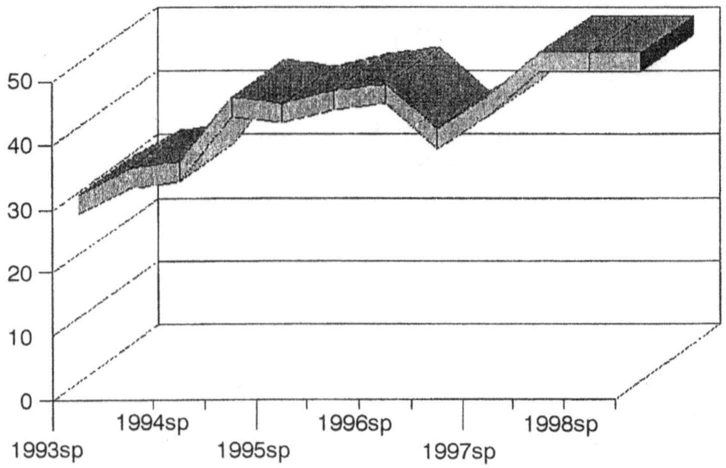

however, provide the background for conducting further experimental study to examine the effectiveness of technology-based teaching and cooperative learning strategies.

TECHNOLOGY-BASED STATISTICS TEACHING MODEL

A technology-based teaching model, as shown in Figure 4, is summarized from my statistics teaching reform and is characterized by its potential to stimulate student statistics learning.

In this model, using technology is the foundation, and using real-

FIGURE 3
Introductory statistics course average scores.

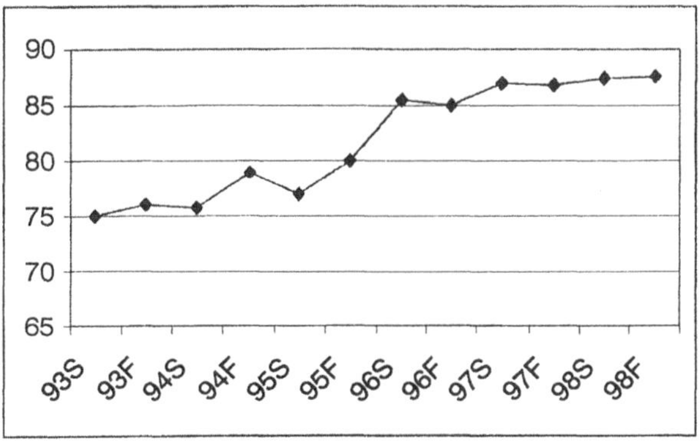

FIGURE 4
Technology-based statistics teaching model.

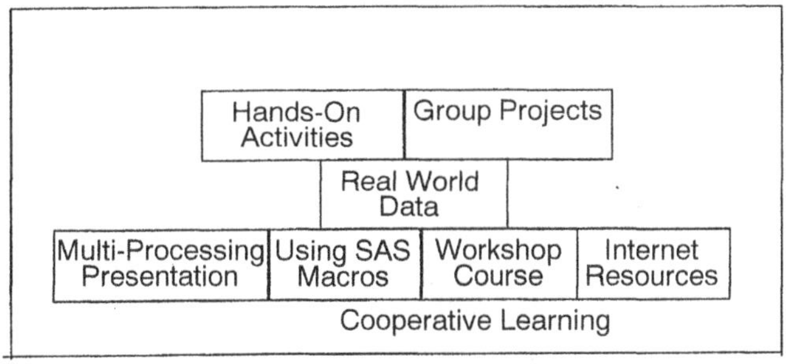

world data is the core of the teaching and applications; both are merged into the process of cooperative learning. This model provides a full structure for technology-based teaching. In further studies, research on the effectiveness of this teaching model may focus on the investigation of each component. It is hoped that our experience and the technology-based model developed therefrom may be helpful for statistics teachers in planning and designing their courses.

REFERENCES

Alavi, M. (1994). Computer-mediated collaborative learning: An empirical evaluation. *MIS, 18*(2), 159-174.

Astlertner, H., & Keller, J. M. (1995). A model for motivationally adaptive computer-assisted instruction. *Journal of Research and Computing in Education, 27*(3), 270-280.

Benswanger, H. (1991). Volition as cognitive self-regulation. *Organizational Behavior and Human Decision processes, 50*(1), 154-178.

Brush, T. A. (1997). The effects on student achievement and attitudes when using integrated learning systems with cooperative pairs. *Educational Technology Research and Development, 45*(1), 51-64.

Christmann, E. P., & Badgett, J. L. (1997). Progressive comparison of the effects of computer-assisted instruction on the academic achievement of secondary students. *Journal of Research on Computing in Education, 29*(4), 325-337.

Cobb, G. (1987). Introductory textbooks: A framework for evaluation. *Journal of the American Statistical Association, 82*, 321-339.

Cobb, G. (1993). Reconsidering statistics education: A national science foundation conference. *Journal of Statistics Education, 1*(1). [On-line]. Available: *http://www.stat.ncsu.edu/info/jse/v1n1/cobb.html*

Dietz, E. J. (1998). The challenges and opportunities of the electronic medium. *Journal of Statistics Education, 52*(4), 469-474.

Fernandez, G. C. J. (1998). Quick and complete results from data analysis: Introductory statistical methods (p. 266). APST270 Introductory Statistics Lab course offered by Department of Applied Economics and Statistics, University of Nevada, Reno, NV 89557.

Fillebrown, S. (1994). Using projects in an elementary statistics course for non-science majors. *Journal of Statistics Education, 2*(2). [On-line]. Available:
http://www.stat.ncsu.edu/info/jse/v2n2/fillebrown.html

Garfield, J. B. (1993). Teaching statistics using small-group cooperative learning. *Journal of Statistics Education, 1*(1). [On-line]. Available: *http://www.stat.ncsu.edu/info/jse/v1n1/garfield.html*

Gatti, G., & Harwell, M. (1998). Advantages of computer programs over power charts for the estimation of power. *Journal of Statistics Education, 6*(3). [On-line]. Available: *http://www.stat.ncsu.edu/info/jse/v6n3/gatto.html*

Gnanadesikan, M., Scheaffer, R. L., Watkins, A. E., & Witmer, J. A. (1997). An activity-based statistics course. *Journal of Statistics Education, 5*(2). [On-line]. Available: *http://www.stat.ncsu.edu/info/jse/v5n2/gnanadesikan.html*

Groeneboom, P., & de Jong, P. (1996). Computer-assisted statistics education at Delft University of Technology. *Journal of Computational and Graphical Statistics, 5*(4), 386-400.

Hogg, R. V. (1972). On statistical education. *The American Statistician, 26*(1), 8-11.

Hogg, R. V. (1990). Statisticians gather to discuss statistics education. *American Mathematics and Statistics News, 169*, 19-20.

Hogg, R.V. (1991). Statistical education: Improvements are badly needed. *The American Statistician, 45*(4), 342-343.

Hunter, W. G. (1977). Some ideas about teaching design of experiments, with 25 examples of experiments conducted by students. *The American Statistician, 31*(1), 12-17.

Keller, J. M. (1983). Motivational design of instruction. In C. M. Reigeluth (Ed.), *Instructional design theories and models: An overview of their current status* (pp. 384-434). Hillsdale, NJ: Lawrence Erlbaum.

Keeler, C., & Steinhorst, R. (1995). Using small groups to promote active learning in the introductory statistics course: A report from the field. *Journal of Statistics Education, 3*(2).

Kellenberger, D. W. (1996). Preservice teachers' perceived computer self-efficacy based on achievement and value beliefs within a motivational framework. *Journal of Research on Computing in Education, 29*(2), 124-140.

Khamis, H. J. (1991). Manual computations–a tool for reinforcing concepts and techniques. *American Statistician, 45*(4), 294-300.

Locke, E. A., & Latham, G. P. (1990). *A theory of goal setting and task performance.* Englewood Cliffs, NJ: Prentice-Hall.

Locke, E. A., & Latham, G. P. (1994). Goal setting theory. In H. F. O'Neil & M. Drillings (Eds.), *Motivation: Theory and research* (pp. 13-29). Hillsdale, NJ: Lawrence Erlbaum.

Liu, L., & Johnson, D. L. (1998). A computer achievement model: Computer attitude and computer achievement. *Computers in the Schools, 14*(3/4), 33-54.

Liu, L., & Fernandez, G. C. J. (1998). Multiple regression and model selection on developing a computer achievement model. In K. LeBouton, T. Cermack, & S. Gupta (Eds.), *Let Your SAS Knowledge Grow* (pp. 228-233). Oakland, CA: Western Users of SAS Software.

Malout, D. B. (1988). The effect of instructional computer games on continuing student motivation. *Journal of Special Education, 21*(4), 27-28.

Moore, T. L., & Roberts, R. A. (1989). Statistics at a liberal arts college. *The American Statistician, 43*(1), 85-90.

Posten, H. O. (1982). Neat data for teaching statistics. *Teaching Statistics, 4*(1), 42-45.

Read, K. L. Q., & Riley, I. S. (1983). Statistics problems with simple numbers. *The American Statistician, 37*(3), 229-231.

Roberts, H. V. (1992). Student-conducted projects in introductory statistics courses. In F. Gordon & S. Gordon (Eds.), *Statistics for the Twenty-First Century* (pp. 109-121). MAA Notes No. 26, Washington, DC: Mathematical Association of America.

Smith, G. (1998). Learning statistics by doing statistics. *Journal of Statistics Education, 6*(3). [On-line]. Available: *http://www.stat.ncsu.edu/info/jse/v6n3/smith.html*

Tabatabai, M., & Gamble, R. (1997). Business statistics education: Content and software in undergraduate business statistics courses. *Journal of Education for Business, 73*(1), 48-54.

Webster, E. H. (1991). Computer augmented instruction in statistics education. *Atlantic Economic Journal, 19*(1), 44.

West, R. W., & Ogden, R. T. (1998). Interactive demonstrations for statistics education on the World Wide Web. *Journal of Statistics Education, 6*(3). [On-line]. Available: *http://www.stat.ncsu.edu/info/jse/v6n3/west.html*

Yilmaz, M. R. (1996). The challenge of teaching statistics to non-specialists. *Journal of Statistics Education, 4*(1). [On-line]. Available: *http://www.stat.ncsu.edu/info/jse/v4n1/yilmaz.html*

Zetterqvist, L. (1997). Statistics for chemistry students: How to make a statistics course useful by focusing on applications. *Journal of Statistics Education, 5*(1). [On-line]. Available: *http://www.stat.ncsu.edu/info/jse/v5n1/zetterqvist.html*

APPENDIX

Internet Resources

I. Animated Java applets and other statistics resources:

1. Statistical Instruction Internet Palette. *http://research.ed.asu.edu/siip*. A palette of resources for statistics students and instructors, including a data gallery, a data web, computing and graphing studios, and an annotated formula gallery. John Behrens, Arizona State University.
2. Collection of Statistics Java Applets. *http://www.isds.duke.edu/sites/java.html*: This collection of Java applets comes from NWP Associates, Inc. Data sets up to 100 rows, 10 columns can be analyzed for free; anything larger requires a subscription.
3. Introductory Statistics Demonstrations. *http://ouvaxa.cats.ohiou.edu/~wallace/class/demonf.html*. Series of animated demonstrations of basic statistics concepts by David Wallace, Ohio University.
4. Internet Guide for Group Projects. *http://www.math.fsc.qut.edu.au/MAB893/manual.html*: A guide for students doing their own data collection and analysis. Includes glossary of statistical terms and brief recipes for using basic statistical techniques. Catriona Hayes and Helen MacGillivray, Queensland University of Technology.
5. Java Applets That Perform a Few Basic Statistical Tests. *http://intrepid.mcs.kent.edu/~blewis/stat/index.html*
6. Internet Glossary of Statistical Terms. *http://www.animatedsoftware.com/statglos/statglos.htm*

7. CTI Statistics Web site. *http://www.stats.gla.ac.uk/cti/* A free on-line statistical resource center. Here you will find software reviews and articles, a searchable guide to software for teaching, a diary of forthcoming statistical events worldwide, a CBL software developers' forum, mailing list information, contact addresses, and links to a wealth of statistical resources worldwide.
8. Graphical analysis page. *http://curriculum.qed.qld.gov.au/kla/eda/intro.htm*

II. Textbook Web sites:

1. A Hyper-Textbook Library to accompany STAT101: Introductory Statistics. Jonathon Coombes, University of Newcastle. *http://surfstat.newcastle.edu.au/surfstat/*
2. Electronic Statistics Textbook. *http://www.statsoftinc.com/textbook/stathome.html* Offers training in the understanding and application of statistics. The material was developed at the StatSoft R&D department based on many years of teaching undergraduate and graduate statistics courses and covers a wide variety of applications, including laboratory research (biomedical, agricultural, etc.), business statistics and forecasting, social science statistics and survey research, data mining, engineering and quality control applications, and many others.
3. UCLA Electronic Textbook *http://www.stat.ucla.edu/textbook/* By the Department of Statistics at the University of California-Los Angeles.
4. HyperStat Oniine. *http://www.ruf.rice.edu/~lane/hyperstat/index.html* A substantial introductory-level statistics book including power, factorial ANOVA, between and within-subjects ANOVA, chi-square and distribution-free tests. Developed by David Lane, Rice University.

III. Web DATA sources:

1. DASL *http://lib.stat.cmu.edu/DASL/* An online library of data files and stories that illustrate the use of basic statistics methods. They provide data from a wide variety of topics so that statistics teachers can find real-world examples that will be interesting to their students. You can use DASL's powerful search engine to locate the story or data file of interest.
2. CHANCE DATA Base: A wealth of materials designed to be helpful in teaching CHANCE Statistics course. CHANCE is an introductory case study course based on current chance events as reported in newspapers and current journals.
3. Journal of Statistics Education Data Archive *http://www.stat.ncsu.edu/info/jse/datasets.index.html*
4. North Carolina State Statistics and Computing Lab Data Sets *http://www.stat.ncsu.edu/sicl/sicldataindex.htm*
5. UCLA Statistics Data Sets *http://www.stat.ucla.edu/data/data.phtml*
6. StatLib–Data Sets Archive *http://lib.stat.cmu*

Carl Lee

Computer-Assisted Approach for Teaching Statistical Concepts

Many students consider statistics to be one of the most difficult and boring subjects of their college experience. The cognitive and non-cognitive factors associated with these attitudes are complex. Non-cognitive factors are related to background, beliefs, and attitudes; whereas, cognitive actors are related to how students learn a subject and to how knowledge is logically transmitted to students using various pedagogical approaches. Causes of non-cognitive factors may have occurred long before a student's college experience. For example, a high percentage of secondary graduates have weak mathematics backgrounds and are intimidated by quantitative reasoning, not to mention the fact that introductory statistics may be the first and only time they encounter the concepts of uncertainty. Adding to the complexity is the traditional lecture and note-taking pedagogy for teaching introductory statistics at the college level. Hogg (1992) said it well:

> Statistics are seldom designed with any idea of what it is that students are supposed to be able to do as a result of having taken the course. . . . Statistics is seen as a 'subject,' rather than a problem solving tool to be used in the scientific method, or a useful way to look at the world around us. (p. 6)

CARL LEE is Professor of Statistics, Department of Mathematics, Central Michigan University, Mt. Pleasant, MI 48859. E-mail: carl.m.lee@cmich.edu

[Haworth co-indexing entry note]: "Computer-Assisted Approach for Teaching Statistical Concepts." Lee, Carl. Co-published simultaneously in *Computers in the Schools* (The Haworth Press, Inc.) Vol. 16, No. 1, 1999, pp. 193-208; and: *Information Technology in Educational Research and Statistics* (ed: Leping Liu, D. LaMont Johnson, and Cleborne D. Maddux) The Haworth Press, Inc., 1999, pp. 193-208. Single or multiple copies of this article are available for a fee from The Haworth Document Delivery Service [1-800-342-9678, 9:00 a.m. - 5:00 p.m. (EST). E-mail address: getinfo@haworthpressinc.com].

The use of computational technology in college introductory statistics is now very common. The simplest and most commonly used computational technology is the scientific calculator, which has been used for decades in introductory statistics to ease the burden of computation. However, calculators cannot help teachers avoid the traditional formulae-driven approach. And, while the advancement of computer technology has led to the development of many user-friendly statistical software packages that allow educators to design new and different ways to teach statistics, these software packages are often designed for data analysis in commercial environments, not for learning statistical reasoning in educational settings. As a result, introductory statistics is often taught using a recipe approach; that is, students learn how to use the software, plug in the numbers, and then try to interpret the computer outputs. Ultimately, statistical concepts and quantitative reasoning skills are not properly taught using the recipe approach. In addition to these computer software applications, graphing calculators and spreadsheets have also been used by instructors (e.g., Bilotti-Aliga, 1998; Boyle, 1998). Furthermore, when considering the use of technology for instruction, one certainly cannot ignore interactive multimedia tools and the World Wide Web. These tools allow students to study the material at their own pace without the limitations of time and location. Clearly, these technology tools will be very beneficial to both instructors and students when they are properly integrated into the instruction. Computer technology is an integral part of the learning process for the computer-assisted approach, from data collection, data analysis, and report writing, to multimedia presentation. In addition, an Internet Web site is used for distributing course-related instructions and materials, such as PowerPoint presentations, projects, activities, and exercises.

Both non-cognitive and cognitive factors are equally important in the process of learning. Little research effort has been done to investigate the non-cognitive factors associated with the learning of introductory statistics (Gal & Ginsburg, 1994; Shaughnessy, 1992). Lee, Meletiou, and Fouladi (1998) conducted an interview study to investigate students' beliefs about, attitudes toward, and relationship with learning styles and motivations. Generally speaking, though, statistics educators have paid overwhelming attention to the cognitive factors when implementing new pedagogies (e.g., Sowey, 1995; Hillmer, 1996; Scheaffer, Gnanadesikan, Watkins, & Witmer, 1996; Giraud, 1997; Rossman & Chance, 1998). The MAA Notes #26 (edited by Gordon &

Gordon, 1992) present a list of interesting and useful ideas for developing innovative pedagogies. *The Journal of Statistical Education* and *The International Statistical Review* are two rich resources for issues related to statistical education. Moore (1997) has presented a thorough discussion on the issues of content, pedagogy, and technology for developing and teaching introductory statistics.

The use of real-world data in an introductory statistics course is based on the assumption that real-world data will motivate students better and provide an environment that is more relevant to their career interests. Stressing real-world applications has greatly improved teaching and learning statistics. However, the fact may have been overlooked that many real-world problems are often beyond college students' experience in introductory statistics. As a consequence, only a small group of students can see the value of these real-world problems and appreciate the statistical applications. A majority of students may continue to hold the same passive attitude and feel disconnected, even though the real-world data are very interesting. Part of the reason may be due to non-cognitive factors associated with learning statistics. First, most students initially have very weak mathematics backgrounds; and, second, they believe that statistics is difficult well before they ever enroll in a statistics class. Third, while real-world data can be very relevant to their future careers, students can foresee neither the connection nor the value. Fourth, most students come to the course with the attitude that all they want to do is to pass the course. This attitude is a difficult obstacle to overcome. Thus, in order to improve students' attitudes and to enhance their technical knowledge, a strategy such as the one presented here needs to simultaneously take into account both non-cognitive and cognitive factors.

In Section 1, the framework for a computer-assisted approach is introduced. Section 2 details each component of the framework. Section 3 addresses the issue of content. Section 4 discusses the procedure of conducting a class lecture using the framework. An illustrative example is provided in Section 5 using both a traditional approach and a computer-assisted, hands-on activity approach. We then conclude with a discussion of the assessment of the computer-assisted approach.

A COMPUTER-ASSISTED APPROACH FOR TEACHING INTRODUCTORY STATISTICS (THE PACE MODEL)

Many educators have pointed out that people learn best by constructing knowledge themselves (e.g., Garfield, 1995; vonGlaserfeld, 1995).

However, as pointed out by Moore (1997), one must be cautioned on how to apply the constructivist learning theory to teaching, especially for difficult concepts and new knowledge that need a great deal of instruction, well-designed guidance, and reinforcement. While the constructivist learning theory is the foundation of the PACE model, because of students' weak mathematics backgrounds and poor attitudes toward the subject, and the difficulty of statistical concepts themselves, we use a more planned and controlled environment in the development of the PACE model.

PACE stands for Projects, Activities, Cooperative learning in a computer classroom environment, and Exercises. The rationale behind the PACE model is based on the following principles:

1. People learn better by constructing knowledge themselves through guided processes.
2. Practice and feedback are essential ingredients for understanding new concepts.
3. Active problem-solving through teamwork promotes active learning.

Rossman and Chance (1998) developed what they called "workshop statistics," which focuses on the hands-on activity component of the PACE model. Some literature has shown that hands-on activities provide a better learning atmosphere, and improve students' attitudes toward learning (e.g., Lee, 1997, 1998a, 1998b; Rossman & Chance, 1998; Witmer, 1998). The use of the cooperative learning strategy for teaching and learning introductory statistics was reported in Keeler and Steinhorst (1995). The PACE strategy attempts to provide a structured framework to integrate projects, hands-on activities that are conducted cooperatively in a computer classroom environment. The paradigm emphasizes:

1. Putting students at the center of the learning process.
2. Having the instructor act as a facilitator who leads and guides students to discover and understand new concepts.
3. Actively engaging students in their learning process.
4. Closely relating statistics as a scientific tool for solving real-world problems.
5. Providing opportunity for students to work as a team.

6. Actively involving students in report writing and oral presentations.

The PACE model integrates the new innovative techniques using modern computer technology and maintains the advantage of the traditional approach of organized structure and reinforcement. This framework carries out the three components of content, pedagogy, and technology addressed in Moore (1997) for an introductory statistics course.

THE BASIC COMPONENTS OF THE PACE MODEL

Component 1: Hands-On Activities

Hands-on activities are the core component for day-to-day classroom activities. A spreadsheet is prepared for each activity, consisting of the purpose for the activity, a procedure for conducting the activity, a worksheet for recording data, and a worksheet for recording the group and class discussion. Each activity is designed to introduce several new concepts and to review previously learned concepts. Students are guided to conduct the activity, collect data, analyze data, and present the findings. Two students work as a team for data analysis, report writing, and presentation. Two or three teams are randomly selected to present their results. The collected data are used for learning various statistical concepts wherever appropriate. For example, "How far are you from home?" is the first activity conducted during the first lecture. This activity is designed to help students discover the relationship between population, samples, selection of measurement, random sampling, graphical presentation, mean, median, standard deviation, percentile, and shape of distribution. Many of these concepts are reinforced in other activities. The concepts and skills learned from an activity are applied in later activities.

Component 2: Self-Selected Projects

The use of self-selected projects is an important component of the PACE model. The concepts of sampling design, experimental design, two independent sample data, paired sample data, categorical data, regression and correlation data are introduced very early on. Students are then instructed to look for two independent sample data, paired sample data, and regression and correlation data, respectively, using any source available to them. The commonly used resources are sports data, weather data, and studies related to the student's own expertise. Self-selected projects, although not as structured, tend to make the learning process more interesting. As a result of using self-selected projects, students have shown greater motivation and interest in learning how statistics can be applied to solve selected problems. Fillebrown (1994) reported a similar experience in the implementation of a self-selected project approach in an elementary statistics course.

It is critical that students' projects are guided throughout the semester. To make the projects work smoothly, guidelines are important.

1. How the project should be conducted.
2. Where to find related information or data sets.
3. How the project will be graded.
4. How a report should be prepared.

Two reports are required. The first is a description of the data source, experimental design, and variables, and a descriptive and graphical analysis of the data. The second report is an analysis of the data using appropriate inferential methods and then drawing conclusions.

Component 3: Cooperative Learning in a Computer Classroom Environment

Teamwork skills are especially needed for college students. Unfortunately, training for such skills is lacking. Part of the reason is the difficulty of planning teamwork exercises in a traditional class lecture setting. Statistical skills are applied in virtually every type of industry. However, they are applied within the context of a larger project. The nature of statistical applications makes teamwork an essential component in the process of problem solving. Unlike the traditional approach, when using hands-on computer activities, it is in fact natural to integrate teamwork into the learning process. There are many different ways to apply cooperative learning (e.g., Keeler & Steinhorst, 1995; Dietz, 1993; Garfield, 1993). Reynolds et al. (1995) presented a thorough discussion of a practical guide to cooperative learning in college mathematics.

In the PACE framework, teamwork is implemented flexibly, depending on the class size and the classroom setup. A typical class size is about 40 students. If a computer classroom is available, the number of computers is usually about 20 to 30. When a hands-on activity is conducted in a computer classroom, two students are assigned as teammates for conducting the hands-on activity, data analysis, report writing, and presentation. During the group discussion, usually four to six groups are formed, based on the setting of the computer classroom. A teammate pair is formed for each activity, while the group discussion team is formed in each class period as needed. Although some literature suggests permanent teams should be formed, each with three to four individuals, our experience indicates that the most frequent complaint about group learning is the difficulty of scheduling time to work together. Therefore, we prefer the flexible setting using two different groupings. The paired team approach provides an opportunity for students to learn how to work together without extra effort to coordinate the time. The in-class group work provides an opportunity for students to express their thinking and to solve problems as a team.

Component 4: Exercises

Regardless of how much an instructor tries to do within the limited time of a class period and how much students discover and understand the concepts, without reinforcement, it is often the case that students will quickly forget what they have learned. We think exercise is an essential tool for retaining concepts and skills. Two exercises are prepared for the PACE framework. One is a simple paper-pencil exercise to reinforce student understanding of individual concepts. The other is a set of problems related to the student's area of study or current events of interest (e.g., the home run chase between Mark McGwire and Sammy Sosa). Students use computers to conduct the analysis and to summarize the findings.

For example, to reinforce the understanding of the shape of distribution, one paper-pencil exercise is to have students find real-world situations where the distribution is skewed to the right, symmetric, and skewed to the left, respectively. A real-world problem for the same purpose can be the salary of university faculty. Students usually are interested in faculty salary. They can analyze the salary using graphical tools and discuss the distribution of salary as a whole, by ranks, and by gender, as well as find the relationship between age and salary.

One can find interesting real-world data on the Internet (e.g., the Data and Story Library at http://www.Stat.cmu.edu/ DASL/).

Component 5: Computing Technology

Computing technology for introductory statistics has evolved from simple calculators to sophisticated software, and interactive multimedia and Web technology. Multimedia and Web technology can be very beneficial for teaching and learning statistical concepts. We currently use these tools to distribute course-related instruction, PowerPoint presentations, and activities for the PACE model. Interactive learning technologies for introductory statistics are becoming available and will be mature in the near future. For example, ActivStats, developed by Velleman (see *www.datadesk.com* for more details) and the Fathom Project (see *www.keypress.com* for more details) are now available. The impact of these interactive multimedia tools has yet to be investigated.

Computing technology is used in every stage of the PACE model. The success of the PACE model relies on the proper integration of computing technology into each component of the model. In order to smoothly integrate computer technology into the process of learning, we suggest that the course be conducted in a computer classroom, where students and the instructor can use the technology at any time. In my case, I conduct the course using a mobile laptop computer lab with 25 laptop computers and a laser printer. The station is Internet ready and uses remote control for each laptop to access the Internet and printer. An appropriate classroom usually houses about 10 large tables with Internet connections. If the mobile laptop technology is not available, one can use a computer classroom setup, or use graphing calculators for activities and a computer lab for projects and reports.

CONTENTS

What content ought to be covered in an introductory statistics course can be a huge debate. The PACE framework chooses to put more emphasis on statistical reasoning and the concepts of variation, and less emphasis on probability and formulae. The order of introducing the content can also be debatable. In our computer-assisted approach, we stress the importance of proper measurement, and introduce the concepts of sampling, experimental design, regression and relationship, and categorical data as early as possible. Various exploratory data analysis tools and their applications are introduced at this stage as well. The importance of variation and shape of the distribution are stressed whenever possible. For example, the first activity, "“How far are you away from home?” quickly leads to the discussion of skewed distribution, because this distribution is usually skewed to the right.

PROCEDURE FOR CONDUCTING THE PACE FRAMEWORK

Before the semester starts, activities are designed with the necessary instructions and worksheets. Early into the semester, students are introduced to various computer software for statistical analysis, report writing, and presentation. The first hands-on activity is also used for demonstrating computer tools. Students begin to conduct self-selected projects after the hands-on activity for regression and correlation is conducted.

Table 1 lists activities that are designed for an introductory statistics course.

Figure 1 provides a basic flow chart for conducting a hands-on activity.

AN EXAMPLE

In this section, the traditional teaching approach and the computer-assisted PACE strategy for the topics of independent samples and paired samples are demonstrated.

The Traditional Approach

A typical lecture for introducing the concepts of two independent sample problems and paired sample problems may be described as follows:

> Two independent sample problems: Two samples are independently chosen at random; one from each population. Let {x1, x2, ... ,xn} represent the observed sample from population A,

and {y1, y2, . . . ym} represent the observed sample from population B.

Paired sample problems: The same variable is measured twice from the same subject before and after a certain treatment. Let {x1, x2, . . . ,xn} represent the observed data before the treatment, and {y1, y2, . . . yn} represent the observed data after the treatment.

Example for two-independent sample problem: Suppose we are interested in comparing pulse rates between male and female students at this university. The group of all male students is population A, and the group of all female students is population B. We will take a random sample of, say, 20 males and observe their pulse rate. Then, we will randomly select, say, 30 females and observe their pulse rates. We will take a random sample independently from each population. This is a two-independent sample problem.

Example for paired-sample problem: Suppose we are interested in the effect of one minute of exercise on the pulse rates among all students at this university. The group of all students at the university is the population. We first randomly select, say, 30 individuals and take their pulse rates. Then, we ask them to do one minute of exercise and take their pulse rates again. Pulse rates are measured from the same individual before and after one minute of exercise. This is a paired sample problem.

Once the data are collected, students need to learn how to analyze it. Let's consider the two independent sample problem using the pulse rate example. To compare the pulse rates between males and females, formulae will be introduced for computing confidence intervals or for testing the hypothesis that the average pulse rate of males is significantly different from females. Students will either use calculators to conduct a step-by-step computation or use a computer to generate output. Students are then taught to interpret the results.

The approach presented emphasizes computation and how to use the formulae. Students passively listen and are busy copying down the

TABLE 1
Computer-Assisted Activities for the PACE Framework

Activity	Intended Concepts Covered
How far are you away from home?	Population, samples, random sampling, measurements, descriptive statistics, graphical tools, shape of distribution, outlier
Can one minute of exercise dramatically increase your heart beat rate?	Two independent samples versus paired samples, distributions, sampling error, factors affecting the measurement, confidence interval, hypothesis testing
How well can palm size predict height?	Regression, correlation, measurement, distributions
If the election is held today, which party do your vote for?	Categorical data, nominal scale, how polls are conducted
Are students customers?	Categorical data, ordinal scale; bias and missing data
How well can you draw a line two inches long?	Bias and variation, replication; within-subject variation versus between-subject variation
Estimate the proportion of each color of M&M candy.	Probability, estimation of proportion, confidence interval, hypothesis testing
Roulette game–what is the chance of winning?	Chance, probability distribution, expected value
How many raisins in each one-half ounce of a box?	Estimation, confidence interval, hypothesis testing
Baseball hitting rates by simulation	Binomial distribution, sampling distribution of sample mean

material and trying to plug numbers into formulae. This approach does not help students understand and be able to design and conduct independent sample designs or paired sample designs in real applications. Students receive the impression that statistics is nothing but formulae and number crunching, and quickly get lost and become bored with the subject. The computation procedures do not connect with the concept of recognizing when to apply the independent sample design and when to apply the paired design, nor do they teach students how to conduct a study or what are the possible sampling errors and factors associated with the pulse rate changes. These concepts are much more important than computational procedures and number crunching.

FIGURE 1
Flow chart for conducting hands-on activities using a PACE model.

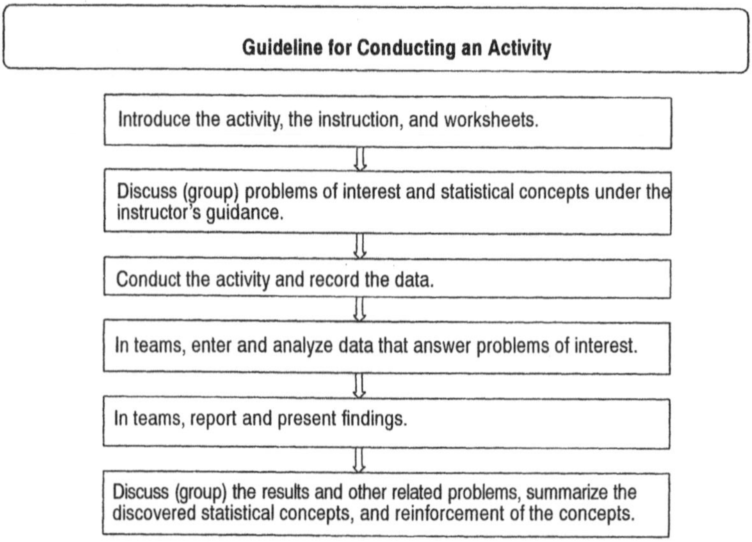

The Computer-Assisted Approach

"Do you think one minute of exercise will dramatically change your pulse rate?" This is our in-class activity. Let me first describe how and why we conduct this activity.

As the worksheet describes, each of you will take your pulse rate for one minute, jump up and down for one minute, and then take your pulse rate again. You will record your own pulse rates and gender in the worksheet, and share with others later. I will give you instructions when to take the pulse rate, when to start exercising, and so on.

In order to make the exercise a fun activity, the instructor may bring a piece of music to class or ask a student to lead the exercise. Once the pulse rates before and after the exercise are taken, have each student announce his/her pulse rates for other students.

After data are collected, it is group discussion time.

You may have noticed that some pulse rates have changed a lot, and some have changed little. Some individuals even have lower pulse rates after exercise. What would be the possible causes of such variation? Your first task in the group discussion is to find out possible factors associated with the pulse rate changes and how to reduce the variation of the measurements.

It would not be surprising that students are able to figure out that factors such as age, weight, gender, whether an individual regularly exercises, and mistakes occurring when counting the rates may be factors associated with the pulse rate changes. A further discussion can be continued to discuss what type of effect each of these factors may have.

The next topic for group discussion is to identify the problems of interest using the collected data, and hypothesize possible results.

Some typical problems of interest identified by groups may include:

1. To compare the rate change before and after the exercise.
2. To compare pulse rates between males and females before the exercise.
3. To compare pulse rates between males and females after the exercise.
4. To compare the rate changes between males and females.
5. If information such as weights, ages, and whether an individual does regular exercise are collected, more sophisticated analysis can be introduced.

The last group discussion before data analysis is to find out the differences among these problems. The concepts of two independent sample problems and paired sample problems can now be introduced for discussion. After this group discussion, students are able to distinguish two independent sample problems and paired sample problems. The computational procedure is then introduced. The concept of rare events is stressed to help students understand the way the decision is made to find out whether the difference in the averages between males and females is large.

Students are paired to work on the data analysis, reporting writing, and presentation of the pulse rate data. Four teams will be randomly selected to present their findings, one for each problem of interest.

Using the computer-assisted approach, students are at the center of the learning process and are guided to discover statistical concepts. Computer technology is essential in data collection, data analysis, report writing and presentation. This example clearly demonstrates the difference between the PACE approach and the traditional approach.

ASSESSMENT OF THE COMPUTER-ASSISTED PACE MODEL

Outcome assessment of students' achievement was conducted using quizzes, tests, projects, and exercises for final grade determination. In addition, a survey was conducted to find out students' satisfaction with the use of activities, projects, cooperative learning set up, and the computer for projects, data analysis, report writing, and presentations. In particular, students were asked to rate the need for using computers for each topic covered in the course. The survey results were reported in Lee (1998b). Generally speaking, virtually every student enjoyed the activities and projects. Computer technology was also rated as a necessary tool for all of the topics except probability distributions and the Central Limit Theorem. The concept of the Central Limit Theorem is perhaps the most difficult concept in introductory statistics. The hands-on activity we chose for learning this concept involves simulation of baseball-hitting averages. More study is needed to investigate and develop better activities for students to learn the concept of the Central Limit Theorem.

An interview study was conducted to assess attitude changes and retention of knowledge two months after taking the course. In order to compare the traditional and PACE approaches, two classes of students were interviewed. One was a traditional class, the other was a computer-assisted PACE class. Nine students from each class were invited for the interview (three with *A*s, three with *B*s, and three with a *C* or lower grade). Some of the findings were presented in Lee, Meletiou, and Fouladi (1998). A clear message from the interview was that students from the computer-assisted PACE class thought that statistics is, in fact, interesting, while students from the traditional class continued to think of statistics as difficult and boring. Some comparisons can be found in Lee (1999). Students were also asked to verbally solve some real problems in the interview. When they were asked about the strategy that they used to solve these problems, the students from the PACE class indicated they used a problem-solving strategy similar to the one learned from the class. The students from the traditional class indicated they looked for a specific formula that they had learned to solve the problem. This distinction clearly shows that the PACE model not only improved attitudes but also improved students' problem-solving skills in solving real-world problems.

REFERENCES

ActivStats (1998). An Interactive Statistical Learning Software. [On-line]. Available: *http://www.datadesk.com/ActivStats/*

Bilotti-Aliaga, M. (1998). Introduction to statistics: Enhancement through the use of graphing calculator. In Lionel Pereira-Mendoza, Kea, L.S., Kee, T.W. & Wong, W.K. (Eds.). *Proceedings of the Fifth International Conference on Teaching Statistics*, (pp. 929-936). Singapore: International Statistical Association.

Boyle, R.G. (1998). PACE2000-Stand alone software package and add-in for Microsoft Excel designed for teaching statistics at introductory-intermediate level. In Lionel Pereira-Mendoza, Kea, L.S., Kee, T.W. & Wong, W.K. (Eds.). *Proceedings of the Fifth International Conference on Teaching Statistics*, (pp. 809-816). Singapore: International Statistical Association.

Data and Story Library (1997). [On-line]. Available: *http://www.stat.cmu.edu/DASL/*

Dietz, E.J. (1993). A cooperative learning activity on the methods of selecting a sample. *American Statistician*, pp. 104-107.

Fathom Project (1998). A dynamic software for learning statistical concepts. [On-line]. Available: *http://www.keypress.com/fathom*

Fillebrown, S. (1994). Using projects in an elementary statistics course for non-science majors. *Journal of Statistical Education*, *Vol. #2*. [On-line]. Available: *http://www.stat.ncsu.edu/info/jse/v2n2/Fillebrown.html*

Gal, I., & Ginsburg, L. (1994). The role of beliefs and attitudes in learning statistics: Towards an assessment framework. *Journal of Statistical Education*, Vol. #2. [On-line]. Available: *http://www2.ncsu.edu/ncsu/pams/stat/info/jse/v2n2/gal.html*

Garfield, J. (1993). Teaching statistics using small-group cooperative learning. *Journal of Statistical Education*, *Vol. #1*. [On-line]. Available: *http://www.stat.ncsu.edu/info/jse/v1n1/garfield.html*

Garfield J. (1995). How students learn statistics. *International Statistical Review*, *63*, 25-34.

Giraud, G. (1997). Cooperative learning and statistics instruction. *Journal of Statistics Education*, *Vol. #5*. [On-line]. Available: *http://www.stat.ncsu.edu/info/jse/v5n3/giraud.html*

von Glaserfeld, E. (1995). A constructivist approach to teaching. In L. Steffe & J. Gale (Eds.), *Constructivism in education*, (pp. 3-16). The University of Georgia, Atlanta: Erlbaum.

Gordon, F., & Gordon, S. (Eds.). (1992). *Statistics for the Twenty-First Century.* MAA (Mathematical Association of America) Notes #26.

Hogg, R.V. (1992). Towards lean and lively courses in statistics. In F. Gordon & S. Gordon (Eds.), *Statistics for the Twenty-First Century,* (pp. 3-13). Washington: Mathematical Association of America, Notes # 26.

Hillmer, S.C. (1996). A problem-solving approach to teaching business statistics. *American Statistician,* 249-256.

Keeler, C., & Steinhorst, R. (1995). Using small groups to promote active learning in the introductory statistics course : A report from the field. *Journal of Statistical Education, Vol. #3.* [On-line]. Available: *http://www.stat.ncsu.edu/info/jse/v3n2/keeler.html*

Lee, C. (1997). Promoting active learning in introductory statistics course using the PACE strategy. In Asela Carlon Monroy (Eds.), *Proceedings of the Fifth International Conference in Mathematics Education,* (pp.199-206). Mexico City, Mexico: Grupo Editorial Iberoamerica, S.A. de C.V.

Lee, C. (1998). An assessment of the PACE strategy for an introductory statistics course. In Lionel Pereira-Mendoza, Kea, L.S., Kee, T.W. & Wong, W.K. (Eds.), *Proceeding of the Fifth International Conference on Teaching Statistics,* (pp. 1215-1221). Singapore: International Statistical Institute.

Lee, C. (1999). A comparison of students' beliefs and attitude towards statistics between technology-rich environment and traditional lecture. In David A. Thomas (Eds.), *Proceedings of the 1999 Math/Science Education Technology (M/SET) Conference,* (pp. 133-138). San Antonio: Association for the Advancement of Computing in Education.

Lee, C., Meletiou, M., & Fouladi (1998). Statistical education theme–Teaching introductory statistics at collegiate level. Paper presented at the *Third Annual Conference of Research in Undergraduate Mathematics Education, South Bend, IN.*

Moore, D. (1997). New pedagogy and new content: The case of statistics. *International Statistics Review, 65,* 123-165.

Reynolds, B., Hagelgans, N., Schwingendorf, K., Vidakovic, D., Dubinsky, E., Shahin, M., & Wimbish, Jr. (1995). *A practical guide to cooperative learning in collegiate mathematics.* MAA (Mathematical Association of America), Notes #37.

Rossman, A., & Chance, B. (1998). Workshop statistics: Dissemination and assessment. In Lionel Pereira-Mendoza, Kea, L.S., Kee, T.W. & Wong, W.K. (Eds.), *Proceedings, the Fifth International Conference on Teaching Statistics,* (pp. 131-138). Singapore: International Statistical Institute.

Scheaffer, R.L., Gnanadesikan, M., Watkins, A. & Witmer, J. (1996). *Activity-based statistics: Instructors resources.* New York:Springer.

Shaughnessy, J.M. (1992). Research in probability and statistics: Reflections and directions. In D.A. Grouws (Eds.), *Handbook on research on mathematics teaching and learning,* (pp. 465-494). New York: Macmillian.

Sowey, E. (1995). Teaching statistics: Making it memorable. *Journal of Statistical Education, Vol. #3.* [On-line]. Available: *http://www.stat.ncsu.edu/info/jse/v3n2/sowey.html*

Witmer, J. (1998). Using activities in Stats 101. In Lionel Pereira-Mendoza, Kea, L.S., Kee, T.W. & Wong, W.K. (Eds.), *Proceedings of the Fifth International Conference on Teaching Statistics,* (pp. 139-144). Singapore: International Statistical Institute.

Edwin Christmann
John Badgett

The Comparative Effectiveness of Various Microcomputer-Based Software Packages on Statistical Achievement

Closely related to mathematics, statistics is an academic discipline whose quality of abstraction presents conceptual problems for numerous students. For example, Garfield and Ahlgren (1988) found that the presentation of statistics to students is often methodologically obscure; and students, in general, inject numbers into prescribed formulae with no understanding of the theoretical bases underlying these formulae, which prohibits the application of previously successful problem-solving skills to new situations. Concurring, Snell & Peterson (1992) contend that the mathematical essence of many statistical formulae exceeds the comprehension of most students enrolled in introductory statistics courses.

Aware of the inherent abstraction of mathematics, Patterson and Smith (1986) asserted that, when computers are properly integrated

EDWIN CHRISTMANN is Associate Professor, Secondary Education, McKay Education Building, Slippery Rock University, Slippery Rock, PA 16057. E-mail: edwin.christmann@sru.edu
JOHN BADGETT is Associate Professor, Secondary Education, McKay Education Building, Slippery Rock University, Slippery Rock, PA 16057.

[Haworth co-indexing entry note]: "The Comparative Effectiveness of Various Microcomputer-Based Software Packages on Statistical Achievement." Christmann, Edwin, and John Badgett. Co-published simultaneously in *Computers in the Schools* (The Haworth Press, Inc.) Vol. 16, No. 1, 1999, pp. 209-220; and: *Information Technology in Educational Research and Statistics* (ed: Leping Liu, D. LaMont Johnson, and Cleborne D. Maddux) The Haworth Press, Inc., 1999, pp. 209-220. Single or multiple copies of this article are available for a fee from The Haworth Document Delivery Service [1-800-342-9678, 9:00 a.m. - 5:00 p.m. (EST). E-mail address: getinfo@haworthpressinc.com].

into mathematics instruction, the students are better able to formulate hypotheses, test them, analyze the results, and refine their mathematical conceptions. In agreement, the National Council of Teachers of Mathematics (NCTM) recommended that students have access to microcomputer-based instruction (NCTM, 1989). Hence, in response to suggestions such as these, many educators are now enhancing their students' understanding of statistical formulae and processes through the graphical displays made possible by computers. More specifically, Christmann and Badgett (1997a) found that microcomputers enhanced the statistical problem-solving skills of both males and females, although the males scored significantly higher than the females, while the females' performance was significantly higher when graphing calculators were used as the medium of instruction. Nevertheless, Snell and Peterson (1992) maintain that computer graphics and simulations are highly conducive to students' (a) discovery of patterns and (b) development of insight into statistical concepts. More precisely, Tijms (1992) cites specific examples of computer graphics that were resultant in successful student understanding of individual statistical concepts; and Janvier, Girandon, and Morand (1993) insist that computers can serve as vehicles for connecting corresponding numerical symbols to students' mental depictions of these symbols. Moreover, Snell and Peterson (1992) report the success of computers in efficiently promoting student understanding of statistical concepts, as do Cunningham (1991), Bishop (1989), Schoenfeld (1990), Rosebery and Rubin (1989), and Hallett (1991).

Obviously, there is an abundance of research supporting the use of microcomputer-based software in the teaching of statistical concepts. However, only a paucity of research has been conducted on software packages that are geared toward the teaching of these concepts. Hence, a primary purpose of this research was to determine the effectiveness of accessible software packages on the statistics achievement of a variety of students through a meta-analytic technique that examines available research targeting the effectiveness of such software packages. Then, another purpose of the study was to determine whether there is a difference between the academic achievement of those statistics students who were exposed to microcomputer-based software, and those who were not. Finally, an attempt was made to determine the presence of a pattern reflecting the effectiveness of microcomputer-based software on statistical achievement over a ten-year period.

PREVIOUS META-ANALYTIC FINDINGS

Kulik and Kulik (1986) found that when exposed to computer-assisted instruction (CAI), college-level students in virtually any demographical educational setting outperformed 60% of their peers who were instructed with traditional methodology, as reflected in a 0.260 meta-analytic effect size. Commensurate with these results, an earlier study by Kulik, Kulik, and Cohen (1980) disclosed an effect size of 0.250. Moreover, Bangert-Drowns, Kulik, and Kulik (1985), using a meta-analytic technique to determine the effectiveness of CAI on the achievement of secondary students, calculated an effect size of 0.260, again showing CAI to be comparatively more effective than traditional instruction.

STATEMENT OF THE PROBLEM

The necessity for improving statistics achievement in American education was the motive for conducting this meta-analysis of all available microcomputer-based software research pertaining to statistics instruction in academic settings. The study has focused on the following research questions:

- Are there differences between the academic achievement levels of statistics students who were exposed to instruction supplemented with microcomputer-based software, and those who received traditional instruction only?
- Are there differences among the effects of seven software packages on students' statistics achievement?
- Is there a pattern of the effectiveness of microcomputer-based statistical software in bolstering statistics achievement over a ten-year period?

A comparison was first made between the effectiveness of microcomputers per se, and traditional instructional methods. Then the research evaluated the comparative effectiveness of seven microcomputer-based software packages on the achievement of statistics students. Finally, the comparative effects of microcomputer-based software on statistics achievement over a ten-year period was determined.

METHODOLOGY

The compiled data from the studies were analyzed through a meta-analysis technique, a secondary statistical analysis or re-analysis of previous research, which is used as a vehicle for answering new questions through existing data (Glass, McGaw, & Smith, 1981). In essence, it is a quantitative application of deduction that would have been impossible through any other previously known methodology (Gall, Borg, & Gall, 1996).

The studies included in this research met the following predetermined criteria:

1. They were conducted on students enrolled in a statistical analysis course.
2. They involved comparisons between the effects of microcomputer-based computer software packages and traditional methodologies on statistics achievement.
3. They had experimental, quasi-experimental, or correlational research designs.
4. The sample sizes had a combined minimum of 20 students in the experimental and control groups.

Nine relevant publications met the predetermined criteria for inclusion in the meta-analysis. Any study that was rejected did not meet all four of these selection criteria. The majority of those publications not meeting the criteria for integration into the study did not statistically analyze the reported data.

ANALYSIS

As previously mentioned, the data were analyzed through a meta-analysis procedure, a technique that relies heavily on the calculation of effect sizes for establishing statistical meaning (Wolf, 1986). According to Glass et al. (1981), effect size is the degree to which a phenomenon is present in the population of the study. In meta-analysis, effect size is calculated to determine the presence of a statistical difference between mean standard deviation units (SD_x) (Wolf, 1986). In the current study, positive effect sizes are indicative of the comparative superiority of microcomputer-based software; whereas, negative effect sizes reflect the comparative superiority of traditional methodology.

META-ANALYSIS

Nine of over 300 studies met the predetermined criteria for inclusion in this meta-analysis. Table 1 presents the author(s), date, type of software, sample size, and the mean effect sizes for the included studies. A total of 703 students participated in the nine studies, resulting in 14 conclusions. The sample size ranged from 25 to 133, with a mean sample size of 50 students.

MEAN EFFECT SIZES

From the 14 effect sizes calculated, an overall mean effect size of the meta-analysis was also calculated. The sum of the 14 effect sizes is 3.587. The mean effect size of 0.256 is positive because those statistics students receiving instruction supplemented with microcomputer-based software attained higher scores than those receiving traditional instruction alone. Cohen (1977) classifies this effect as small.

Wolf's (1986) interpretation of average effect in SD_X units for the comparison between the traditional instruction group and the microcomputer-based software group indicates that the average student exposed to statistical software showed academic achievement that exceeded that of the students who were exposed to traditional instruction. Moreover, Figure 1 displays that the typical student moved from the 50th percentile to the 60th percentile when exposed to microcomputer-based software.

Effectiveness of Various Microcomputer-Based Software Packages

A comparison of the mean effect sizes of various software types is shown in Table 2. The 14 effect sizes were categorized into three software types (CAI, problem-solving, and statistical software packages). The mean effect size was calculated for each software type where statistics achievement and the effects of the software were studied. The mean effect size of problem solving software on statistics

TABLE 1
Author, Date, Software, Type, Sample Size, and Effect Sizes of Various Computerized Software Packages

Author(s)	Date	Software	Type	*n*	ES
Athey, S.	1987	SAM	CAI/Mentoring	25	0.360
Christmann, et al.	1997	MyStat	Statistical Software	50	0.208
Dorn, M.	1993	Hypercard	Problem-Solving	35	1.150
Dorn, M.	1993	Hypercard	Problem-Solving	39	0.708
Gilligan, W.	1990	Minitab	Statistical Software	49	−0.075
Gilligan, W.	1990	Minitab	Statistical Software	29	0.093
Gratz, et al.	1993	SPSS/PC	Statistical Software	55	0.538
Gratz, et al.	1993	SPSS/OC	Statistical Software	55	−0.555
Gratz, et al.	1993	SPSS/PC	Statistical Software	55	−0.037
Hollowell, et al.	1991	TruSTAT	Statistical Software	133	0.087
Porter, et al.	1996	Statistical Exercises	CAI	39	0.942
Rosen, et al.	1994	SPSS/PC	Statistical Software	25	0.178
Song, P.W.	1992	MyStat	Statistical Software	57	−0.122
Song, P.W.	1992	MyStat	Statistical Software	57	0.112

achievement was the largest effect size in the meta-analysis. The smallest mean effect size was calculated on statistical software packages (Mystat, Minitab, and SPSS). In descending order, the mean effect sizes according to software types are: 0.929, 0.651, and 0.043, respectively (see Table 2).

FIGURE 1
Overall average effect size in SD_x units.

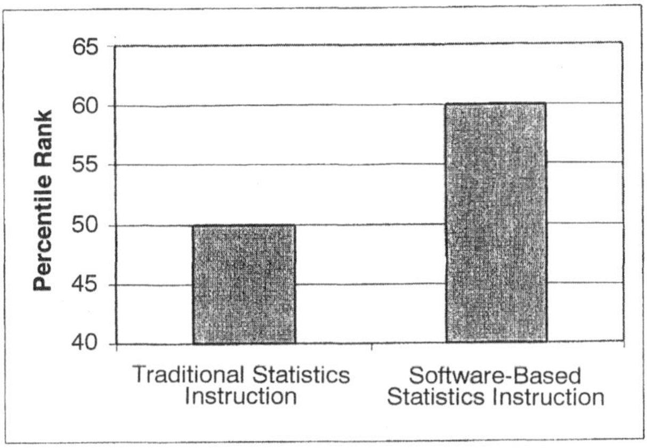

TABLE 2
Mean Effect Sizes of Software Types

Type of Software	*n*	ES
CAI	2	0.651
Problem-Solving	2	0.929
Statistical Software	10	0.043

Relationship Between Effect Sizes and Time

A correlation of the data revealed that the mean effect sizes and progressive time span in years were not significantly related, -0.052, $n = 7$, $p > .05$, one tail (Table 3). Furthermore, Figure 2 displays a scatterplot showing a low, negative correlation between consecutive years and the effectiveness of microcomputer-based statistical software.

CONCLUSIONS AND RECOMMENDATIONS

In accordance with the previously stated objectives, the study has assessed the effectiveness of microcomputer-based statistical software

TABLE 3
Mean Effect Sizes of Software Types

Year	Number of ES	Mean ES
1987	2	0.360
1990	2	0.009
1991	1	0.087
1992	2	−0.005
1993	5	0.361
1994	1	0.178
1997	1	0.208

FIGURE 2
Relationship between mean effect sizes and years.

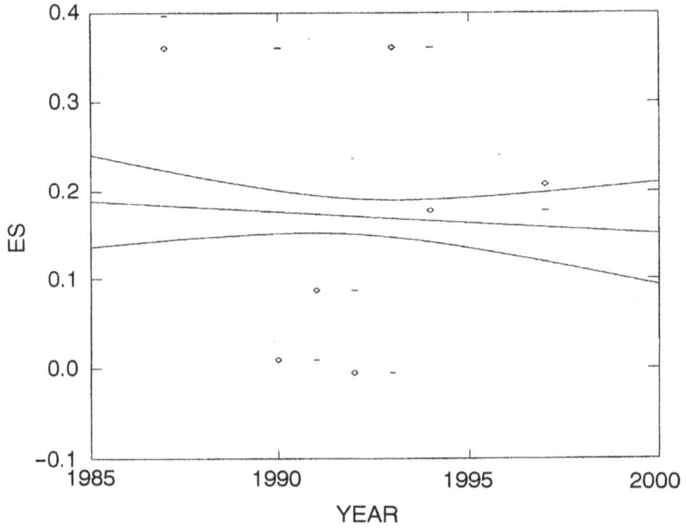

on academic achievement. These objectives were to determine: (a) the effectiveness of accessible software packages on statistics achievement; (b) whether there is a difference between the academic achievement of those statistics students who were exposed to microcomputer-based software and those who were not; (c) the presence of a pattern of the effectiveness of microcomputer-based software on statistics achievement over a 10-year period.

Effect sizes. Cohen (1977) has provided the following ranges for effect size interpretations: *ES* = 0.200 to 0.499 (small effect); *ES* = 0.500 to 0.799 (medium effect); and *ES* = 0.800 and above (large effect). Tallmadge (1977) reports that an effect size difference of 0.250 or more is considered to be educationally significant. Effect sizes (*ES*x) in standard deviation units (SD_X) show the degree of overlap between control and experimental groups (see Figure 1). This representation depicts which percentage of the experimental group exceeds the upper half of the control group. Moreover, this technique clearly demonstrates the impact of microcomputer-based statistical software on academic achievement.

Mean effect sizes. The mean effect size calculation across the 14 conclusions of the nine studies generated by this meta-analysis is 0.256. Cohen (1977) interprets this effect size as small. The difference in academic achievement resulting from microcomputer-based statistical software was an improvement of over 10.0 percentile ranks, from the 50th to the 60th, with respect to the central region of the distribution. Using this measure, it can be concluded that microcomputer-based statistical software is somewhat more effective than traditional methods of statistics instruction in raising overall academic percentile rankings among statistics students. This finding supports the research of Kulik et al. (1986), which shows a positive effect of computer-based education on academic achievement among college students. However, Kulik et al. (1986) calculated a lower average effect size, 0.260 (ironically, the same effect size that was calculated on the 1985 Bangert-Drowns et al. study on secondary education students), which demonstrates that those students exposed to computer-based education (CBE) improved their academic achievement from the 50th to the 60th percentile. In comparison to the average effect size in the current meta-analysis, the research of Kulik et al. (1980) indicates that the academic achievement of students exposed to microcomputer-based statistical software is about the same as that of those students exposed to computer-based education in Kulik's earlier study. As indicated by a 0.929 effect size for problem-solving and a 0.651 effect size for CAI, these packages appear more effective than statistical software, which has an effect size of 0.043. Hence, the statistical software packages seem obviously less effective than the other two modes of instruction.

The low negative correlation of -0.052 between mean effect size and years indicates a very slight but, nevertheless, lessening of the effectiveness of microcomputer-based software package on statistics achievement over a ten-year period; but, as previously mentioned, very little research has been conducted on the comparative effects of the various software packages on statistics achievement, and only a smattering of research has been directed toward the progressive effects of these packages on statistics achievement. Hence, it is blatantly obvious that additional research needs to target these areas.

Reliability and Literature Searches

Publication biases may have increased the effect sizes disclosed in this meta-analysis, in that many professional journals have elected to publish those articles that are favorable to technology. Moreover, certain methods of statistical analysis produce results that do not necessarily reflect actual effect sizes (Wolf, 1986). Nevertheless, Bangert-Drowns, et al. (1985) have identified two distinct purposes of meta-analysis: (a) to describe the current research literature, and (b) to attempt to uncover the "true" effect sizes of a treatment. This meta-analysis focuses on the first purpose, describing the span of research results to date. However, determining the true effect sizes of treatments is very difficult, partially because of the necessity of reporting means and standard deviations. Nevertheless, Cohen (1977) recommends that effect sizes appear in every report.

CONCLUSIONS

Kerr (1996) cautions that "the preconception about technology's efficiency and effectiveness is strong, and often leads to directions that ultimately turn out to be unproductive (p. xiv)." Obviously, microcomputer-based statistical software provides students with a tool that enhances academic achievement. However, Christmann et al. (1997a) remind educators and researchers that meta-analysis is simply a method of analyzing research; therefore, no meta-analytic study can accurately forecast future developments in statistics achievement. Perhaps, a partial answer to our problems in American education is the integration of computers into college classrooms. However, the -0.052 correlation between mean effect size and years seems somewhat congruent with the -0.762 correlation that Christmann et al. (1997b) found in a progressive comparison between CAI and traditional instruction over a 12-year period. Hence, these figures definitely reflect the necessity for continuing research regarding the effectiveness–or lack of effectiveness–of computers in American scholastic and collegiate institutions. Undoubtedly, this additional research is imperative for determining the most effective means for improving statistics achievement in all academic settings.

REFERENCES

References marked with an asterisk indicate studies included in the meta-analysis.

*Athey, S. (1987). A mentoring system incorporating expertise to guide and teach statistical decision making. (Doctoral dissertation, The University of Arizona, 1987). *Dissertation Abstracts International, 46*, 02-A.

Bangert-Drowns, R.A., Kulik, C.C., & Kulik, J.A. (1985). Computer-based education in secondary schools. *Journal of Computer-Based Instruction, 12*(3), 59-68.

Bishop, A. (1989). Review of research on visualization in mathematics education. *Focus on Learning Problems in Mathematics, 11*, 7-16.

*Christmann, E.P., & Badgett, J.L. (1997a). The effects of computers and calculators on male and female statistics achievement. *Journal of Computing in Higher Education, 9*(1), 49-58.

Christmann, E.P., Badgett, J.L., & Lucking, R.A. (1997b). Progressive comparison of the effects of computer-assisted instruction on the academic achievement of secondary students. *Journal of Research on Computing in Education, 29*(4), 325-337.

Cohen, J. (1977). *Statistical power analysis for the behavioral sciences*. New York: Academic Press.

Cunningham, S. (1991). The visualization environment for mathematics education. In W. Zimmerman & S. Cunningham (Eds.). *Visualization in teaching and learning mathematics* (pp. 67-76). Washington, DC: Mathematical Association of America.

*Dorn, M.J. (1993). The effect of an interactive, problem-based hypercard modular instruction on statistical reasoning (Doctoral dissertation, Southern Illinois University at Carbondale, 1994). *Dissertation Abstracts International, 55*, 09-A.

Gall, M.D., Borg, W.R., & Gall, J.P. (1996). *Educational research: An introduction*. New York: Longman.

Garfield, J., & Ahlgren, A. (1988). Difficulties in learning basic concepts in probability and statistics: Implications for research. *Journal for Research in Mathematics Education, 19*, 44-63.

*Gilligan, W.P. (1990) The use of a computer statistical package in teaching a unit of descriptive statistics. (Doctoral dissertation, Boston University, 1990) *Dissertation Abstracts International, 51*, 07-A.

Glass, G. McGaw, B., & Smith, M.L. (1981). *Meta-analysis in social research*. Beverly Hills: Sage.

*Gratz, Z.S., Volpe, G.D., & Kind, B.M. (1993). Attitudes and achievement in introductory psychological statistics classes: Traditional versus computer supported instruction. Paper presented at the annual meeting of Teachers of Psychology, Ellenville, New York.

Hallett, D.H. (1991). Visualization and calculus reform. In W. Zimmerman & S. Cunningham (Eds), *Visualization in teaching and learning mathematics* (pp. 121-126). Washington, DC: Mathematical Association of America.

*Hollowell, K.A., & Duch, B.J. (1991). *Functions and statistics with computers at the college level.* Paper presented at the annual meeting of the American Educational Research Association, Chicago, IL.

Janvier, C., Girardon, C., & Morand, J. (1993). Mathematical symbols and representations. In P.S. Wilson (Ed.), *Research ideas for the classroom: High school mathematics* (pp. 79-102). New York: Macmillan.

Kerr, S. (1996). Editor's Preface. *Technology and the Future of Schooling: National Society for the Study of Education, 95*(2), xi-xiv.

Kulik, J.A., Kulik, C.C., & Cohen, P.A. (1980). Effectiveness of computer-based college teaching: A meta-analysis of findings. *Review of Educational Research, 50*(4), 525-544.

Kulik, C.C., & Kulik, J.A. (1986). Effectiveness of computer-based education in colleges. *AEDS Journal, 19*(2/3), 81-108.

National Council of Teachers of Mathematics. (1989). *Curriculum and evaluation standards for school mathematics.* Reston, VA: Author.

Patterson, J. & Smith, M. (1986). The role of computers in higher-order thinking. *Microcomputers and Education: National Society for the Study of Education, 85*(1), 81-108.

*Porter, T.S., & Riley, T.M. (1996). The effectiveness of computer exercises in introductory statistics. *Journal of Economic Education, 27*(1), 291-299.

Rosebery, A., & Rubin, A. (1989). Reasoning under uncertainty: Developing statistical reasoning. *Journal of Mathematical Behavior, 8*, 205-219.

*Rosen, E.F., Feeney, B., & Petty, L.C. (1994). An introductory statistics class and examination using SPSS/PC. *Behavior Research Methods, Instruments, & Computers, 26*(2), 242-244.

Schoenfeld, A. (Ed.). (1990). *A source book for college mathematics teaching.* Washington, DC: Mathematical Association of America.

Snell, J.L., & Peterson, W.P. (1992). Does the computer help us understand statistics? In F. Gordon & S. Gordon (Eds.), *Statistics for the twenty-first century* (pp. 167-188). Mathematical Association of America Notes Number 26. Washington, DC: Mathematical Association of America.

*Song, P.W. (1992). The effect of using computer software and computer simulations to teach statistics and probability in a Korean college (Doctoral dissertation, The University of Iowa, 1992). *Dissertation Abstracts International, 53*, 07-A.

Tallmadge, G. (1977). *The joint dissemination review panel book.* Washington, DC: National Institute of Education and the U.S. Office of Education.

Tijms, H. (1992). Exploring probability and statistics using computer graphics. In F. Gordon & S. Gordon (Eds.), *Statistics for the twenty-first century* (pp. 189-197). Mathematical Association of America Notes Number 26. Washington, DC: Mathematical Association of America.

Wolf, F. (1986). *Meta-analysis: Quantitative methods for research synthesis.* Newbury Park, London: Sage.

Christina Dokter
Larry Heimann

A Web Site as a Tool for Learning Statistics

Reforms in the way professors teach statistics have been steadily growing. Non-math students such as social science and business students often have a very difficult time learning statistics. This problem stems partly from a lack of interest in math, as well as, from their negative attitudes toward statistics. Social science students often view statistics as the hardest course in college. The need to have social science students approach statistics with positive attitudes and have them see that statistical methods are applicable to everyday life (Hogg, 1991; Yilmaz, 1996) have generated many new approaches to teaching statistics. Chance (1997), Dietz (1993), Fillebrown (1994), Gnanadesikan (1997), Ledolter (1995), Mackisack (1994), and Snee (1993) have tried new ways to replace traditional lecturing with hands-on activities. In their studies, students tend to show new interest in statistics and become more eager to learn the subject. Similarly, Fillebrown (1994), Roberts (1992), and Smith (1998) used project-based curricula that improved both students' attitudes and academic performance.

Others (Ferrall, 1995; Ferris & Hardaway, 1994; Gatti & Harwell,

CHRISTINA DOKTER is a Doctoral Candidate, College of Education, Michigan State University, East Lansing, MI 48823. E-mail: dokterch@pilot.msu.edu
LARRY HEIMANN is Associate Professor of Political Science, Department of Social and Decision Sciences, Carnegie Mellon University, 319E Porter Hall, Pittsburgh, PA 15213. E-mail: lheimann@andrew.cmu.edu

[Haworth co-indexing entry note]: "A Web Site as a Tool for Learning Statistics." Dokter, Christina, and Larry Heimann. Co-published simultaneously in *Computers in the Schools* (The Haworth Press, Inc.) Vol. 16, No. 1, 1999, pp. 221-229; and: *Information Technology in Educational Research and Statistics* (ed: Leping Liu, D. LaMont Johnson, and Cleborne D. Maddux) The Haworth Press, Inc., 1999, pp. 221-229. Single or multiple copies of this article are available for a fee from The Haworth Document Delivery Service [1-800-342-9678, 9:00 a.m. - 5:00 p.m. (EST). E-mail address: getinfo@haworthpressinc.com].

1998) have tried computer programs as learning aids to help students better understand statistical reasoning. They used software that helped individual students further grasp statistical concepts they may not have understood in class.

In all of the above, the trend has been a move away from textbook-based curricula to a more constructivist, situative learning model of teaching statistics.

Our project is an attempt at combining real-life statistical application problems within a database of the Internet as a tool for facilitating communication and learning. While this project does not remove the textbook from the classroom, the attempt is to apply more real-life statistical problems to enhance what is taught in the text. As such, a Web-based instructional site will help students grasp statistical concepts they may not readily understand. The underlying theory behind such an approach is mainly social constructivism. We define *social constructivism* along ideas implicated in Vygotsky's philosophy of learning and development. As such, we encourage cooperative learning among students through dialogue and the process of argument and negotiation. Through such processes, students reach a state of equilibrium in their thinking and learning, expanding their Zone of Proximal Development.

For our purposes, we will limit our definition of social engagement to the joint activity implications underlying socio-historic theory found in Vygotsky's ideas. Vygotsky's writings imply a socio-historic theory that says learning becomes internalized through mediation that occurs during the actual problem-solving or shared task with others (Cole et al., 1978, pp. 56-57).

In this case, mediation is defined as student's interaction with the Web site at *http://www.msu.edu/~heimann/PS801*. The tools of the Web site, its contents, and the people interacting within, provide a way of reconstructing the cognitive process. They use online discussions, threaded discussions, and online problem solving to reconstruct their understandings of statistical concepts. Meaning is constructed through social mediation, negotiation, and dialogue (Newman, Griffin, & Cole, 1989).

Others have approached a noncomputer-oriented social constructivist view with mixed results. Garfield (1993) and Smith (1998) suggest that cooperative learning enables students to share knowledge and learning within group settings.

Several studies of technology-based collaborative learning have assessed test and survey results showing higher levels of perceived learning and skill development than students who engaged in group learning in conventional contexts (e.g., Hiltz, 1994; Harasim, Hiltz, Teles, & Turoff, 1995; Alavi, 1994; Shneiderman, 1994). Although they were conducted in classes that dealt with complex learning, these studies were not research about statistics classes. Our project attempts to combine statistics learning with technology as a socially mediated tool. As one of the first in such an attempt, we wanted to design a Web site specifically to meet the needs of a political science statistics class, and then analyze the limitations and cautions such a project entails.

THE PROBLEMS STUDENTS HAVE LEARNING STATISTICS

When political science students try to solve statistics problems, there are two major problems they often face. The first is, "How do I get the machine to give me the results that I want?" The "machine" is often a computer, but for many it is only a scientific calculator. Many social science students do not like math and have only an elementary foundation in the subject. They are also often not familiar with computational technology.

Even if they are familiar with computers and calculators, students must overcome another, often more difficult problem as they learn statistics. That difficulty is one of problem identification. Since there are many different techniques available to use in statistics, students often do not know which to use. This is particularly confusing when the teacher demonstrates that there are multiple ways of solving a problem–all of which lead to the same answer. Precisely because statistics is as much an art as it is a science, students need practice in recognizing which methods and approaches are most applicable to a particular problem. They also need training in converting a story problem into something they can handle as statisticians.

To do this, students must first be able to classify the problem and place it in one of the major categories of statistics problems, and second, be able to decompose the problem so that they can extract all relevant information. Following that, they must be able to execute the appropriate method(s) to find the solution. This last step is often the easiest; the first two steps cause students the most problem.

Textbooks are often less than helpful in dealing with the first two steps because exercises are often placed at the end (or within) the chapters that cover the relevant material. The result is that the student who does only the textbook problems rarely gets much insight into step one, since the problems at the end of the binomial distribution chapter probably can be classified as binomial distribution problems. In addition, textbook problems often give students exactly the information they need to solve the problem in an easily digestible format, hence delimiting their practice on step two–extracting all relevant information. In contrast, some of the exam problems are more reflective of the real world because there is either irrelevant information included or subtlety in the way the information is presented. What textbook problems are good at is helping students with the mechanics of executing the appropriate method(s) to find the solution (step three). Not surprisingly, students often state that once they work a problem past the first two steps, they can figure it out from there.

Because the issue of problem identification is difficult to address using traditional methods and affects a larger percentage of students, it is this problem that needs to be addressed first. The problem of students' unfamiliarity with "machine" usage will naturally diminish with time as they are involved in more computer-based learning.

THE WEB SITE FOR LEARNING STATISTICS

According to the needs of the class, the developers set up the following sites:

1. *Virtual Study Center.* This center is composed of eWeb, a virtual learning center and the hub of social constructivist learning activities. Readers can sign on to this site as a guest and read about how eWeb was created with social constructivist principles.

We entered multiple-choice exercises into the testing/exercises function of eWeb. This provides a databank of practice problem sets that are in mixed order of the textbook chapters. That way, students can practice problem identification and receive immediate feedback. Three sets of such tests were entered so that students could practice problems before the finals. When students complete an exam, they are able to check the answers immediately by going to the answer section. This section gives the right answer with a short explanation.

Students can use the Forum and the Chat Room in this site to conduct online discussions about questions and problems that occurred in class or in the textbook exercises. The purpose of this site is to foster social constructivist learning by using communication technologies over the Internet. Our theory is that, as students use the social interactive components of eWeb, such as the Forum and the Chat Room, their knowledge changes through the process of challenge and negotiation.

The professor and the teaching assistant used the Chat Room by setting up a designated hour each week in which they would log on and students could ask them questions online. The Forum is a threaded discussion section: A student can enter a comment and then have other students read and reply to his or her comment.

Such virtual tools enhance learning without having the students physically come together. The main benefactors are commuting students who might otherwise be prevented from joining study groups.

2. *Testing Center.* This center archives old exams, class-assigned problem sets, and solutions to textbook exercises. The main purpose of this center is to equalize and legitimize the distribution of old exams that are usually passed from friend to friend in inner circles. Students who are not a part of such a circulation can now access the old exams through this site. Students can access the tests online or download them to their computer.
3. *Surfing Center.* The Surfing Center contains World Wide Web information and links related to statistics and political science so that students can expand their understanding from various cyber resources. In addition, when the professor linked his favorite sites, including a Star Trek link, the serious mood of simply learning statistics was broken.
4. *Teaching Center.* The Teaching Center contains explanations about statistics concepts and related topics with color and explicit graphics and diagrams. Topics include distributions, nonparameter inferences, and inference relationships. This center enhances student understanding of topics discussed in class.
5. *Document Center.* The Document Center contains a syllabus with class office hours, course requirements, and meeting times. Although students receive a syllabus in class, such a site maintains updates and changes, and the professor does not have to

worry about students who misplace syllabi or question the changes in schedule.

THE BENEFITS OF THE WEB SITE

Given that students have problems applying statistical concepts to real-life problems, and that textbooks do little to solve this problem, the Web site has been set up to create opportunities for students to engage in various forms of statistical problem solving. Through Web-based practice quizzes, dialogue with colleagues and the professor, and through the many study aid choices provided by the Web site, we have provided a means for students to expand their resources for learning statistics.

As students use the Web site, we evaluate its effectiveness as a study aid for the class. Initially, the students seemed reluctant to go to the Web site. However, with time, we expect them to become more comfortable using this type of computer technology. The result will be circular. The more they use the site, the more they will become familiar with how the site can help them, and thus the more they will use the Web site. We also expect student motivation to increase as they become more familiar with the site. The difficulty students have in identifying and distinguishing various types of statistics problems should be mitigated as students begin to use the testing site more often.

Through the Web site, students have developed an informal support network to help others through difficult subjects like statistical analysis. Part of this support network includes access to files that contain all previous exam questions and solutions. Because these files exist and cannot be controlled, students have bountiful opportunity to quiz themselves and discuss problems with one another.

The Web site also revealed some information about using technology that unexpectedly changed the dynamics of the teacher-pupil interaction. The professor was able to take advantage of the Chat Room by setting up Chat Room hours in lieu of office hours. As such, the professor said, "My teaching assistant and I log on to the Chat Room at designated times four times each week. This is great for students who live far away and cannot come to my office during designated office hours. Also, if another student asks me a similar or the same question as was discussed on the eWeb, I can tell them to go back and check out the transcript dialogue files in the Chat Room."

This is an example of technology adaptation that is purely utilitarian in nature. The class passes through three stages. Both the students and the professor first try to figure out the mechanics of using the Web site. Then, they begin to adapt to the technology to suit their needs. Lastly, they begin looking for other ways to maximize this technology. We expect the professor to adapt other functions of the Web site to better suit his teaching needs as he realizes more of its potential. For example, the testing and exercises site may eventually coincide with actual class lectures.

Such findings suggest that further evaluations should concentrate on qualitative approaches that look at the process of learning with technology rather than just looking at the outcomes of learning.

TECHNOLOGY AND LIMITATIONS

As developers, we found some limitations with using Internet technology, first because the technology itself is in its infancy, and second, because many school computer labs are not yet fully equipped to accommodate what Net-based technology can offer. Below are examples that illustrate this point:

1. Ideally, we would like a program that asks a statistics question and then provides a wide, blank space for students to work out the solutions. Such a program would allow students to enter the statistical solutions into the answer bank so that others can see their process of solving the problems.

Entering statistics symbols into a Web site is a function that is still in its infancy in html-based environments. Thus, the best way to enter math symbols is by typing them in a word-processing program, copying and pasting them into a graphics program, and then saving them as a .gif or .jpg file. The final step is then to paste them into the correct designated answer slot on the Web page. This means that all students in the class need to know how to type math symbols, load this text into a graphics program, save it as a file compatible with html, and load it onto a Web page.

In our project, the questions were entered manually in multiple-choice form along with immediate feedback. We were able to type in the feedback by typing symbols out in alphanumeric form. For example, "Your answer is incorrect; the correct answer is c, $2\pi \times .004$; please study normal distribution in the teaching center or the text."

2. The developers wanted to create a video of the professor explaining how to use SPSS and load it onto the Web site for all students to access. The problem is that, even if we use video streaming and our computers could accommodate a large file, our students use machines that are not compatible with this type of technology, and the students may not know how to download plugins for sound and video. Sometimes, developers have to suppress their desire to use the latest technology and add the latest tool that is available. A simple videotape would do the same thing.
3. The developers found that simplicity was the best lure for students. Web site menus had to be easy to understand, the design had to be uncluttered, and students needed initial guidance on how to use the Web site, even though, to the developers, everything seemed self-explanatory.

As technology develops at an unprecedented rate, we expect similar projects to become more facilitative. White boards and video conferencing should, for example, overcome the problem of incompatibility statistical symbols with html. Extensive qualitative research is needed as technology grows and provides better methods for aiding the teaching and learning of statistics.

REFERENCES

Alavi, M. (1994). Computer-mediated collaborative learning: An empirical evaluation. *MIS Quarterly 18*(2), 159-174.

Chance, B. L. (1997). Experiences with authentic assessment techniques in an introductory statistics course. *Journal of Statistics Education*, *5*(3). [Online]. Available: *http://www.stat.ncsu.edu/info/jse/v5n3/chance.html*

Cole, M. Et Al. (1978). *Mind in Society*. Cambridge, MA: Harvard University Press.

Dietz, E. J. (1993). A cooperative learning activity on methods of selecting a sample. *The American Statistician*, *47*, 104-108.

Ferrall, C. (1995). Interactive statistics tutorials in strata. *Journal of Statistics Education*, *3*(3). [Online]. Available: *http://www.stat.ncsu.edu/info/jse/v3n3/ferrall.html*

Ferris M., & Hardaway, D. (1994). Teacher 2000: A new tool for multimedia teaching of introductory business statistics. *Journal of Statistics Education*, *2*(1). [Online]. Available: *http://www.stat.ncsu.edu/info/jse/v2n1/ferris.html*

Fillebrown, S. (1994). Using projects in an elementary statistics course for non-science majors. *Journal of Statistics Education*, *2*(2). [Online]. Available: *http://www.stat.ncsu.edu/info/jse/v2n2/fillebrown.html*

Garfield, J. (1993). Teaching statistics using small-group cooperative learning. *Journal of Statistics Education*, *1*(1). [Online]. Available: *http://www.stat.ncsu.edu/info/jse/v1n1/garfield.html*

Gatti, G., & Harwell, M. (1998). Advantages of computer programs over power charts for the estimation of power. *Journal of Statistics Education*, *6*(3). [Online]. Available: *http://www.stat.ncsu.edu/info/jse/v6n3/gatti.html*

Gnanadesikan, M. (1997). An activity-based statistics course. *Journal of Statistics Education*, *5*(2). [Online]. Available: *http://www.stat.ncsu.edu/info/jse/v5n2/gnanadesikan.html*

Harasim L., Hiltz, S.R., Teles, L., & Turoff, M. (1995). *Learning networks: A field guide to teaching and learning online*. Cambridge, MA: MIT Press.

Hiltz, S. R. (1994). *The virtual classroom: Learning without limits via computer networks: A field guide to teaching and learning online*. Cambridge, MA: MIT Press.

Hogg, R. V. (1991). Statistical education: Improvements are badly needed. *The American Statistician*, *45*(4), 342-343.

Ledolter, J. (1995). Projects in introductory statistics courses. *The American Statistician*, *49*, 364-367.

Mackisack, M. (1994). What is the use of experiments conducted by statistics students? *Journal of Statistics Education*, *2*(1). [Online]. Available: *http://www.stat.ncsu/edu/info/jse/v2n1/mackisack/html*

Newman, D., Griffin, P., & Cole, M. (1989). *The construction zone: Working for cognitive change in school*. Cambridge University Press.

Roberts, H.V. (1992). Student-conducted projects in introductory statistics courses. In F. Gordon & S. Gordon (Eds.), *Statistics for the twenty-first century*. Mathematical Association of America, No. 26, 109-121. Washington, DC: Mathematical Association of America.

Smith, G. (1998). Learning statistics by doing statistics. [Online]. Available: *http://gsmith.pomona.edu/Econ57 paper.html*

Snee, R. D. (1993). What's missing in statistical education? *The American Statistician*, *47*, 149-154.

Shneiderman, B. (1994). Education by engagement and construction: Can distance learning be better than face-to face? [Online]. Available: *http://www.hitl.washington.edu/scivw/EVE/distance.html*

Index

Access (database software), 71
Ahlgren, A., 209
Airasian, P.W., 7,16
Alavi, M., 177,223
Altrichter, H., 103
American Psychological Association (APA). See APA
Analysis, data
  qualitative. See Qualitative data analysis
  quantitative. See Quantitative data analysis
  statistical software. See Statistical software
Anderson, G.E., 84
Anderson, P., 54
APA, 71
Apostle, L., 54,67,68,69
Approach Software Corporation, 71
AQUAD (statistical software), 126-127
ARL, 33
Armstrong, B., 55
Aspects (collaborative writing software), 46-47
Association of Research Libraries (ARL). See ARL
Astlertner, H., 177
Atkinson, P., 35-38,146
ATLAS/ti (statistical software), 3,40,43,136-143,150
Atomistic qualitative data analysis, 21,45,51,118,121-123, 145-146,147f

Babbie, E.R., 58
Bader, C.R., 68
Badgett, J., 176,209,210
Bakeman, R., 126
Ballistic Computing, 48
Bangert-Drowns, R.A., 218
Barry, C., 146
Becker, B.J., 54
Becker, H.J., 81
Beckley, R.F., 68
Beebe, L., 54
Beichner, R.J., 82
BEON, 105
Berger, C.F., 84
Berger, R.M., 53,54
Berlin, Technical University of, 136
Bertrande, D., 68
Bible, King James version, 39
Bilotti-Algia, M., 194
Bishop, A., 210
Bleich, H.L., 68
Bloom, B., 86,87
Bloom's Taxonomy, 86,87f
Blyler, N., 54
BMD (statistical software), 3,36
Bodrova, E., 14-15
Bohm, G., 93
Boles, K., 82
Borg, W.R., 212
Boyle, R.G., 194
Bozeman, W.C., 153
Bristol On-Line Educational Network (BEON). See BEON
British Telecom PhC, 101-114
Brown, C.C., 77
Brush, T.A., 177

CAI. See Computer-assisted approaches to teaching statistics
Calvin, W.H., 68

Campbell, D.T., 58
CAQDAS (statistical software), 37-40,150
Chance, B.L., 194,196,221
Charney, D., 54
Chase, S., 42
Chenail, R., 25-26
Christrnann, E.P., 176,209,210, 218-219
Claris-Page (hypertext creation software), 32
Clark, R.E., 153
Clements, D.H., 13-14
Clouser, J., 55
CMC, 22-23
Cobb, G., 173,178
Cochran-Smith, M., 84
Code-and-retrieve data analysis, 37-40,122-123,135-136, 146-149. See also Qualitative data analysis
Code-A-Text (discourse analysis software), 145
Coffey, A., 35-38,146
Cohen, J., 213,217,218
Cohen, P.A., 211
Cole, M., 222
Collaborative Writing Engine (collaborative writing software), 47-48
Collaborative writing software
  Aspects, 46-47
  Collaborative Writing Engine, 47-48
  CommonSpace, 46-47
  Daedalus Integrating Writing Environment, 46-47
  Norton Textra, 46-47
Collins, P., 39-43,45,50,118,120,122
CommonSpace (collaborative writing software), 46-47
Computer assisted instruction (CAI). See Computer-assisted approached to teaching statistics
Computer Assisted Qualitative Data Analysis Software (CAQDAS). See CAQDAS (statistical software)
Computer programs for data analysis. See Statistical software
Computer-assisted approaches to teaching statistics, 151-171, 173-191,193-208,209-220, 221-229
Computer-mediated communication (CMC). See CMC
Constructivism
  defined, 5-6, 16-19
  historical perspectives, 6-14
  modern, 14-16
  Piagetian theories, 8-14
  quantitative research vs. qualitative research, 5-6
  teaching statistics. See Teaching statistics
  Vygotskian theories, 8-14,222-223
Content analysis software, 145
Corbin, J., 118
Council on European Social Science Data Archives (Web site), 29
Creative writing software
  FinalDraft, 50
  Movie Magic Dramatica, 48-49
  Script Wrzard, 50
  ScriptThing, 50
  ScriptWare, 50
  SidebySide, 50
  StoryVision, 49
  Writer's Block, 49
Cricket Graph (graphics software), 156-157
Cummings, R., 5
Cunningham, S., 210

Daedalus Group, 46-47
Daedalus Integrated Writing Environment (collaborative writing software), 46-47
Data analysis

atomistic. See Atomistic qualitative data
code-and-retrieve. See Code-and-retrieve data analysis. See also Qualitative data analysis
holistic. See Holistic qualitative data analysis
statistical software. See Statistical software
qualitative data analysis. See Qualitative data analysis
Data archives
Council on European Social Science Archive (Web site), 29
Fianish Social Science Data Archive (Web Site), 29
Harvard-MIT Data Center (Web site), 29
Qualidata (Web site), 29
Data collection using computers, 29-30
Database management software
Access, 71
benefits of, 79
Dbase, 71
described, 55,70-79
FOLIO, 145
FoxPro, 71
Paradox, 71
quality of research findings using an EMBD, 53-65
Davis, C.S., 61
Davis, N.E., 101,104,107,113
Dbase (database management software), 71
de Jong, P., 176
Delta principle, 82
Description qualitative data analysis, 119
Development and research approach vs. research and development approach, 101-114
Dietz, E.J., 174,175, 198,221
Dillon, P., 101
Discourse analysis software, 145
Document maintenance
cards files, 54-55,67-68
database management software, 55,70-79
personal filing systems, 67-68
Dokter, C., 221
Dowie, C.A.I., 104
Dyslexia,17

Editorial management database (EMBD). See EMBD
Eisner, E.W., 6
Electronic discussion groups, 22-23
Electronic Journal Alliance, 33
Electronic publications, 32-34
EMBD, 54-64
Ethnograph (statistical software), 3,37-38,43-45,117,123-129, 137-139,143-150
Exemplars, lists of, 120
Ezop, P., 55

Fernandaz, G.C.J., 173,174,176,177, 178
Ferrall, C., 221-222
Ferris, M., 221-222
Feyerabend, P., 120
Fielding, N., 38
Fillebrown, S., 198,221
FinalDraft (creative writing software), 50
Finnish Social Science Data Archive (Web site), 29
FOLIO (database software), 145
FoxPro (database software), 71
Frankenstein monster, 35
Friday, J., 119
FrontPage (hypertext creation software), 32

Gal, I., 194
Galford, R.R., 68

Gall, J.P., 212
Gall, M.D., 212
Gamble, R., 174,178
Garfield, J.B., 175,176,195,198, 209,222
Gatti, G., 176,221-222
Gay, L.R., 56,57,77
General theory, 118,142
Gentry, C.G., 68
Ginsberg, L., 194
Giradon, C., 210
Giraud, G., 194
Glass, G., 212
Gnanadesikan, M., 176,194,221
Gold, S., 47
Gordon, F., 194-195
Gordon, S., 194-195
Gottman, J.M., 126
Graham, A., 47
Graphics software, 156-157
Griffin, P., 222
Groeneboom, P., 176
Group Logic, 46-47
Groupware. See Collaborative writing software

Hall, R.R., 47
Hallett, D.H., 210
Hanley, S., 13-14,16
Harasim, L., 223
Hardaway, D., 221-222
Harrington, S.A., 53,67
Hart, D., 10
Harvard-MIT Data Center (Web site), 29
Harwell, M., 176,221-222
Haycock, A.B., 68
Heimann, L., 221
Hermeneutics, 119-120,122
Herson, J., 46
Hillmer, S.C., 194
Hiltz, S.R., 223
Hogg, R.V., 174,178,193,221
Hoke, F., 77
Holbrook, B., 35-38,146
Holistic qualitative data analysis, 21,39,43,45-51,118,121-123, 142,145,147f,148
Holmes, T.F., 68
Huberman, M., 36,118
Hutchins, R.D., 68
Hypercard (statistical software), 213-219
Hypertext (creation software)
  Claris-Page, 32
  FrontPage, 32
  PageMill, 32

ICT
  benefits of, 112-114
  described, 101-102
  development and research approach, 102-105
  ISDN applications, case studies, 105-109
  NGFL, 109-110
  U.K. Deparrment of Employment Learning Methods Branch, 101-114
  U.K. Virtual Teacher Centre, ICT Educational Research Centre, 101-114
ICT Educational Research Centre, U.K., Virtual Teacher Centre, 101-114
Information and communications technologies (ICT). See ICT
Instructional technology
  ICT, 101-114. See also ICT
  research statistics course, 151-171
  teacher action research, 81-98
Internet
  Internet resources. See Web sites
  Internet search engines, 26-28
  Web-based model for teaching statistics, 221-229
Interviewing, 40-43
ISDN applications, case studies, 105-109
Ishler, M., 7

Jahoda, G., 68
Janvier, C., 210
Jaramillo, J.A., 7,12-13
Jick, H., 55
Johnson, D.L., 1,53,174,175
Johnson, R.C., 55
Jost, M., 21,117

Kauffman, J.M., 15,18
Keeler, C., 196,198
Kelle, U., 35,36,38-40,118,119,120, 146-147
Kellenberger, D.W., 175
Keller, J.M., 175,176,177
Kelley, A., 85
Kerlin, B., 25
Kerr, S., 218
Khamis, H.J., 176,178
Kimmance, J.K., 104
Kindsvatter, R., 7
King James Bible, 39
Kirt, S.A., 53,54
Koch, G.G., 61
Kohlberg, L., 8,10
Kozma, R.B., 153
Kozulin, A., 15
Kuczynski, L. 126
Kulik, C.C., 211,217,218
Kulik, J.A., 211,217,218
Kusch, J., 93
Kutner, M.H., 61
Kwalitan (statistical software), 3,38-39

Latham, G.P., 175
Learning theories
  Piagetian, 8-14
  Vytgotskian, 8-14,222-223
Ledolter, J., 221
Lee, C., 193,194,196,206
Lee, R., 38
Leong, D.J., 14-15
Lipsett, F.R., 68
Lists of exemplars, 120
Liu, L., 1,53,173,174,175
Local theory, 118-119,142
Locke, E.A., 175
Lonkila, M., 122
Lytle, S.L., 84

Mackisack, M., 221
Maddux, C.D., 5,81
Maintaining research documents
  card files, 67-68
  database management software, 70-79
  personal filing systems, 67-70
Malout, D.B., 177
Mantgani, A., 55
Mason, J., 146
McGaw, B., 212
Meletiou, M., 194,206
Microcomputer-based software for teaching statistics, 209-220. See also Statistical software
Miles, M., 36,118
Minitab (statistical software), 3
Models, teaching
  PACE, 195-207
  technology-based, 173-191
  Web-based, 224-228
Moore, D., 195,196,197
Moore, T.L., 178
Morand, J., 210
Moursand, D., 82
Movie Magic Dramatica (creative writing software), 48-49
Mulder, J., 38,43-45,122
Murphy, E., 13-14
MyStat (statistical software), 3

Nachtscheim, C.J., 61
National Council of Teachers of Mathematics (NCTM). See NCTM
National Science Teachers Association, 6,210
NCSA Emerging Technologies, 47

NCTM, 6,210
Neter, J., 61
Newman, D., 222
NGFL,109-112
Non numerical Unstructured Data by simple but powerful processes of Index Searching and Theory-building (statistical software). See NUD*IST (statistical software)
Non-controversial uses of computers, 21-34
Norris, J., 24,150
Norton Textra (collaborative writing software), 46-47
Nova Southeastern University, School of Social and Systemic Studies, 25
NUD*IST (statistical software), 3,38-39,43-45,117,130-136, 137-141,143-150

O'Donnell, J., 33
Ogden, R.T., 174,177
Okerson, A., 33
Orrnrod, J.E., 7

PACE model for teaching statistics
  assessment of, 206-207
  components, 197-200
  described, 195
  procedures, 201-206
  statistical course content, 200-201
PageMill (hypertext creation software), 32
Palumbo, D.B., 153
Pan, A.C., 81,93
Paradox (database software), 71
Patterson, J., 209
Peshkin, A., 6
Peterson, W.P., 209,210
Piaget, J., 8-19
Piagetian theories, 8-14
Poland, R., 110-111
Posch, P., 103
Posten, H.O., 178
Postpositivist quantitative research, 35-36,118
PowerPoint (presentation software), 157-158
Presentation software, 157-158
Programs, software. See Software
Projects, Activities, Cooperative learning in a classroom environment, and Exercises (PACE) model. See PACE model
Publishing research results
  problems associated with, 53-64
  using computers, 30-34

Qualidata (Web site), 29
Qualitative data analysis
  AQUAD (statistical software), 126-127
  ATLAS/ti (statistical software), 3,40,43,136-143,150
  atomistic. See Atomistic qualitative data analysis
  benefits of computer usage, 51
  BMD (statistical software), 3,36
  CAQDAS (statistical software), 37-40,150
  code-and-retrieve data analysis, 37-40,122-123,135-136, 146-149
  computers and, 21-51,117-150
  content analysis software, 145
  Ethnograph (statistical software), 3,37-38,43-45,117,123-130, 137-139,143-150
  historical perspectives of computer usage, 35-45
  holistic. See Holistic qualitative data analysis
  Hypercard (statistical software), 213-219

limiting factors defined using an EMBD, 54-64
MyStat (statistical software), 3,213-219
non-controversial uses of computers for, 21-22
NUD*IST (statistical software), 3,38-39,43-45,117,130-136, 137-141,143-150
qualitative research. See Qualitative research
SAM (statistical software), 213-219
SAS (statistical software), 3,36
SPSS statistical software, 3,36,143,213-219,288
Statistical software, 117-150,213-219. See also Statistical software
TruSTAT (statistical software), 3,213-219
word processing and, 43-45. See also Word processing software
Minitab (statistical software), 3,218-219
Qualitative Inquiry Project (Web site), 25-26
Qualitative research
blending with quantitative research, 2
CMC, 22-23
constructivism. See Constructivism
data analysis. See Qualitative data analysis
data collection using computers, 29-30
description of, 21,119
document maintenance, 67-80
electronic discussion groups, 22-23
electronic publications, 32-34
general theory, 118,142
hermeneutics, 119-120,122
Internet search engines, 26-28
interviewing, 40-43
local theory, 118-119,142
non-controversial uses of computers in, 21-34
publishing results using computers, 30-34
Qualidata (Web site), 29
Qualitative Inquiry Project (Web site), 25-26
Qualitative Research Page (Web site), 25, 150
Qualitative Research Resources on the Internet (Web site), 25-26
quality of research. See Quality of research
QualPage (Web site), 24-25,150
research, defined, 53
statistical software, 117. See also Statistical software
storytelling, 41-42,120-122,148
verstechen, 119-120,124
vs. quantitative research, 5-6
Web sites, 23-26,28-29,150
Qualitative Research Page (Web site), 25,150
Qualitative Research Resources on the Internet (Web site), 25-26
Quality of research
limiting factors defined using an EMBD, 54-64
problems associated with findings and conclusions, 58-59
problems associated with measurements and instruments, 58
problems associated with references, 59
problems associated with research design, 57-58
problems associated with theme and focus, 56-57
problems associated with writing, 59-60
QualPage (Web site), 24-25,150
Quantitative data analysis
limiting factors defined using an EMBD, 54-64

statistical software. See Statistical software
Quantitative research
blending with qualitative research, 2
constructivism. See Constructivism
Council on European Social Science Archive (Web site), 29
data archives, 28-29
Finnish Social Science Data Archive (Web site), 29
Harvard-MIT Data Center (Web site), 29
postpositivist quantitative research, 35-36,118
quality of research. See Quality of research
research, defined, 53
statistical software. See Statistical software
vs. qualitative research, 5-6
Web sites, 28-29

Read, K.L.Q., 178
Reed, W.M., 151,153
Research
blending qualitative and quantitative research, 2
changing tools and techniques, 1-3
defined, 53
documents, maintaining using database management software, 67-80
purposes of, 118-121
qualitative data analysis. See Qualitative data analysis
qualitative research. See Qualitative research
quality of. See Quality of research
quantitative data analysis. See Quantitative data analysis
quantitative research. See Quantitative research
technology-based teacher action research, 81-98
Reynolds, B., 198
Reynolds, T., 55
Riel, M., 81
Riley, I.S., 178
Ringel, M., 68,74
Roberts, H.V., 221
Roberts, R.A., 178
Roper Organization, 29
Rosebery, A., 210
Rossman, A., 194,196
Rowel, P.P., 68
Rubin, A., 210
Rundall, T., 54

Sagor, R., 84
SAM (statistical software), 213-219
Sands, P., 46
Sanger, J., 54
SAS (statistical software), 3,36
Schaeffer, R.L., 176,194
Schoenfeld, A., 210
Schwartz, J.E., 82
Script Wizard (creative writing software), 50
Scriptdude, 48
ScriptThing (creative writing software), 50
ScriptWare (creative writing software), 50
Shaughnessy, J.M., 194
Shermis, M.D., 84
Shneiderman, B., 223
SidebySide (creative writing software), 50
Singer, K., 67,68,74
Sixth Floor Media, 46-48
Skinner, B.F., 34
Slavin, R.E., 7
Smeltzer, L.R., 54
Smith, G., 174,221,222
Smith, J., 35,118,120
Smith, M.L., 209,212
Snee, R.D., 221

Snell, J.L., 209,210
Society for Information Technology in Education, 6
Software
  collaborative writing. See Collaborative writing software
  content analysis. See Content analysis software
  creative writing. See Creative writing software
  database management. See Database management software
  discourse analysis. See Discourse analysis software
  graphics. See Graphics software
  hypertext creation. See Hypertext creation software
  microcomputer-based packages for teaching statistics, 209-220
  presentation. See Presentation software
  statistical. See Statistical software
  word processing. See Word processing software
Somekh, B., 103,113
Soper, M.E., 54,64,68
Sowey, E., 194
SPSS (statistical software), 3,36,143,228
Spuck, D.W., 153
Stanley, J.C., 58
Statistical Exercises (statistical software), 213-219
Statistical software
  AQUAD, 126-127
  ATLAS/ti, 3,40,43,136-145
  benefits of, 117,148-149
  BMD, 3,36
  CAQDAS, 37-40,150
  described, 117-118
  disadvantages of, 148-149
  Ethnograph, 3,37-38,43-45,117, 123-130,137-139,143-149
  Hypercard, 213-219
  Kwalitan, 3,38-39
  Minitab, 3,213-219
  MyStat, 3,213-219
  NUD*IST, 3,38-39,43-45,117, 130-136,137-141,143-150
  QualPage, 150
  SAM, 213-219
  SAS, 3,36
  SPSS, 3,36,143,213-219,228
  Statistical Exercises, 213-219
  TtuSTAT, 3,213-219
Statistics, teaching of. See Teaching statistics
Stein, H., 93
Steinhorst, R., 176,196,198
Stephenson, C., 83
Stermmer, P.M., 84
Stigleman, S., 77
Still, M., 107
Stokes, M.E., 61
Storytelling, 41-42,120-122,148
StoryVision (creative writing software), 49
Strauss, A., 118
Suter, W.N., 53

Tabatabai, M., 174,178
Taxonomy, Bloom's, 86,87f
TBAR model, 85-98
Teacher action research, technology-based
  benefits and limitations of, 94-98
  Bloom's taxonomy, 86,87f
  description of, 81-85
Teaching statistics
  cognitive vs. non-cognitive factors, 193-195
  computer-assisted approaches, 151-171,173-191,193-208, 209-220,221-229
  constructivism. See Constructivism
  effectiveness of microcomputer-based software packages, 213-219
  historical perspectives, 173-174

Internet resources, 190-191
limiting factors, 161,223-224
methods of, 151-171,175-178
motivational factors, 174-175,178-179
PACE model, 195-207
reforming classroom teaching methods, 151-152,179-188, 209-211,221-224
technology-based teaching model, 173-191
to instructional technology graduate students, 151-171
using microcomputer-based software packages, 209-220
Web-based teaching model, 224-228
Zone of Proximal Development, 222
Technical University of Berlin, 136
Technology-based action research (TBAR) model. See TBAR model
Technology-based model for teaching statistics, 173-191
Technology-based teacher action research
benefits and limitations of, 94-98
Bloom's taxonomy, 86,87f
description of, 81-85
TBAR model, 85-98
Telemantics Centre, University of Exeter, 101
Teles, L., 223
Thesaurus
Eric Descriptors, 74-75
Psychological Index Terms, 74-75
Theories of learning. See Learning theories
Thomas, G.F., 54
Thornton, P., 101,107,110-111
Thorogood, M., 55
Tijms, H., 210
Trend, R., 101,106
Troen, V., 82
TruSTAT (statistical software), 3,213-219
Tucker, N., 46,55
Turoff, M., 223

U.K. Department of Employment Learning Methods Branch, 101-114
U.K. National Grid for Learning NGFL. See NGFL
U.K. Virtual Teacher Centre, ICT Educational Research Centre, 101-114
Universities
Nova Southeastern University, School of Social and Systemic Studies, 25
Technical University of Berlin, 136
University of Alberta, 24
University of Essex, Department of Sociology, 29
University of Exeter, Telemantics Centre, 101
University of Missouri, Rolla, 47
University of Pennsylvania, 33
University of Wales, School of Social and Administrative Studies, 37
Yale University, 33
Utterback, N., 68

Valsiner, J., 14-15
Van der Veer, R., 14-15
Verstehen, 119-120,124
VISE, 136-145. See also ATLAS/ti (statistical software)
Visualization, Integration, Serendipity, and Exploration (VISE). See VISE
von Glaserfeld, E., 195
Vygotsky, L.S., 8-19,222-223
Vytgotskian theories, 8-14,222-223

Walley, T., 55
Walsh, M.E., 7,16
Wanat, C., 68,69
Wasserman, W., 61
Watkins, A., 176,194
Web sites
  Council on European Social Science Archive, 29
  Finnish Social Science Data Archive, 29
  Harvard-MIT Data Center, 29
  Qualitative Inquiry Project, 25-26
  Qualitative Research Page, 25-26,150
  Qualitative Research Resources on the Internet, 25-26
  qualitative research sites, 23-24
  QualPage, 24-25,150
  Qualidata, 29
  resources for teaching statistics, 190-191,221-229
Web-based model for teaching statistics, 221-229
Webster, E.H., 176
Wertsch, J.V., 8,10
West, R.W., 174,177
Wilen, S., 7
Willis, J., 21,117
Witmer, J., 176,194,196
Wolf, F., 212,213,218
Word (word processing software), 30-31,44,145
Word processing software
  use in qualitative data analysis, 43-45
  Word, 30-31,44,145
  WordPerfect, 30-31,145
WordPerfect (word processing software), 30-31,145
Wright, B., 104,107
Wright, K., 69
Writer's Block (creative writing software), 49
Writers Computer Store, 48

Yale University, 33
Yilmaz, M.R., 174,178,221

Zbikowski, J.M., 93
Zehr, M.A., 83,89
Zetterqvist, L., 174
Zone of Proximal Development, 222